Understanding GIS

GIS

An ArcGIS Project Workbook
Second Edition

Christian Harder | **Tim Ormsby** | **Thomas Balstrøm**

Published by Esri Press, Redlands, California

Foreword

More than twenty years ago, I was involved with a book project here at Esri. We produced and published a software workbook called *Understanding GIS: The ARC/INFO Method*. We were just a group of people at this little GIS software company in Redlands, California, who saw that the users of our Arc/Info software needed better guidance on how to use these tools for real analysis projects. Our original goal had been to write a book called *Getting Started with ArcInfo*. More than 500 pages later, we realized that we had described a methodology for doing GIS projects and needed to change the title.

The key idea was to set up a problem in the book, provide the data, then let the students work through the entire process. Little did we know at the time that our book would become a worldwide bestseller and inspire a generation of technically savvy geographers.

Fast forward to 2013. GIS has evolved to keep pace with a changing technology landscape. There have been huge innovations in geospatial data models and user interfaces; GIS now lives and breathes on the web. Much has changed. Yet the need for our users to understand how to organize and think about a GIS project persists. This book, *Understanding GIS: An ArcGIS Project Workbook,* revives the spirit and method of the original that inspired it. I'm pleased to say that the new edition carries the torch while also integrating the new paradigm. It is my hope that it will inspire a new generation of GIS professionals and practitioners.

Clint Brown
Director of Software Products
Esri, Redlands, California

Table of contents

Preface

What's new in the second edition?

This second edition of *Understanding GIS: An ArcGIS Project Workbook* has been completely revised and tested to be compatible with versions 10.1 and 10.2 of ArcGIS for Desktop. New graphics have been created using the Windows 7 operating system. Some steps have changed to reflect changes in how the software works. Lesson 9 (available online as was the case in the first edition) has been completely rewritten to reflect the latest advancements in the rapidly-evolving ArcGIS Online environment. The demographic data has been updated with 2012 census data. The final results of the analysis are the same as they were in the first edition, although some intermediate results may differ slightly. Numerous small mistakes that came to our attention have been fixed.

Background

In the late 1980s, Esri published a software workbook called *Understanding GIS: The ARC/INFO Method*. This book was the first to offer a practical, project-oriented introduction to a commercial GIS software product, and it became popular as both a self-study tutorial for working professionals and as a lab manual in college classrooms.

ArcGIS 10.1 for Desktop is different from the Arc/Info software for which the original book was written. Computers and operating systems are different, too. Ongoing development by Esri has extended the software's capabilities and reworked its architecture to keep pace with advances in technology. Not the least of these advances is the sharing of geographic data and applications on the web. When the original book was published, the World Wide Web was still a couple of years in the future. In addition to new GIS functionality, there have also been significant changes to GIS data models, data storage, and user interfaces. All these changes notwithstanding, the underlying geographic approach (what was then called "The Arc/Info Method") remains basically the same. That approach can be summarized as follows: frame the problem, explore the study area, prepare the data, perform the analysis, and present the results. The same method is followed in this new book.

This book isn't a new edition of the original—the original is part of software history. This is a new book from start to finish. It draws inspiration from the original, but incorporates the latest developments to ArcGIS and reflects the world in which GIS is practiced today.

Description

Understanding GIS: An ArcGIS Project Workbook is a tutorial designed around a multifaceted problem: that of finding a suitable location for a park by the Los Angeles River in Los Angeles, California. Using real spatial data (from the City of Los Angeles and other providers) and realistic requirements, you'll complete all the essential phases of a GIS analysis project from planning, to execution, to follow-up.

Goals

One goal of this book is to help you become a proficient user of ArcGIS 10.1 for Desktop software. Another goal, ultimately more important, is to teach the geographic approach to problem-solving. GIS students are often frustrated because software tools are presented in contexts that don't make clear how the tools relate to one another or how they serve larger purposes. This book incorporates software functionality in a meaningful process of analytic thinking.

Of course, GIS has many uses that are not analytical—data management and cartography to name a couple. Even within the general category of analysis, there are different kinds of problems requiring different tools and strategies. This book is not all things GIS. It's not all things ArcGIS, either. There are many aspects of the software that aren't relevant to our problem and aren't explored in the book. Nevertheless, the book covers a lot of ground. When you're done, you should have a good working knowledge of the software and a strong sense of how to use it productively on your own.

Audience

People come to GIS with varying background knowledge and experience. We loosely classify this book as intended for "ambitious beginners." If you have no prior experience with GIS or any of its wellsprings (geography, cartography, earth science, and computer database technology among them), you may find it more challenging than someone with exposure to these areas. On the other hand, thanks to GPS and household Internet mapping applications, location-based technology is becoming ever more familiar. Many things that needed explanation a few years ago have now merged into the background of common technical savvy.

This book is mainly written for:

- College-level or graduate students in GIS-related disciplines who want to learn ArcGIS 10.1 for Desktop software and GIS practice.
- College and university teachers who want a classroom lab manual to supplement their GIS instruction.
- New GIS professionals who want to strengthen or expand their knowledge of ArcGIS 10.1 for Desktop software.
- Professionals in other technical fields who want cross-training in GIS.
- Current ArcGIS for Desktop users transitioning from earlier software versions to version 10.1 or above.

There are no prerequisites for the book, and we've tried to write it so it will work for anyone with a serious interest in learning GIS. During the course of development and review, we found that many experienced users found things of value in it, too. If you're brand new to GIS and are using the book outside a classroom or support group, be sure to take full advantage of the book's web site—especially the GIS TV videos.

Structure

As mentioned, the book follows a single project, so it's best to work through it from start to finish. Lesson results are provided in the event that you can't do every lesson or run into trouble.

The book has nine lessons, each divided into two or more exercises of varying length. Lesson 9 is only introduced in the book: the exercises for this lesson need to be downloaded from the Understanding GIS web site.

Exercises are fully scripted, so you won't encounter gaps in instruction, but frequently repeated operations aren't spelled out in detail every time. For example, after you've clicked a toolbar button a few times, we won't keep showing pictures of it. Likewise, after you've opened a dialog box a few times, we won't keep telling you how to open it.

Conceptual information is supplied as needed, either directly in the text or in call-out boxes. A few complex topics have a page to themselves. Figures, mostly screen captures, are common throughout the book. Their main purpose is to show you the correct state of your software at a particular point. Some figures are annotated to emphasize settings that need attention or to help interpret a result.

Appendix A discusses assumptions and complications lying behind the analysis. Appendix B is the data license agreement. The book's index will help you find your way back to important concepts, operations, and tools.

The exercises were created on computers running the Microsoft® Windows 7 operating system. They should be compatible on other Windows operating systems, although there may be small differences in operating system paths and the like. All images that are screen captures of software reflect a "Windows Classic" configuration.

Resources

Understanding GIS web site. The book's online support site is the Understanding GIS web site:

http://resources.arcgis.com/en/communities/understanding-gis

Come here to watch videos of the exercises, access our forum, get lesson results, check for workarounds to problems, and more.

Lesson results. The book's exercises are cumulative, with the results of one exercise defining the starting point of the next. For this reason, your results at the end of every lesson have to be right. The lessons include many screen images as visual confirmations of progress, so by the end of an exercise, you should know whether you got the correct results. If you did, you can carry them forward. If you have problems, or if you skip an exercise, you can get result data and maps for any lesson. Results are available at the Understanding GIS web site.

Additional exercises. The exercise instructions for Lesson 9, "Share results online," need to be downloaded from the Understanding GIS web site.

Videos. Every exercise has a narrated how-to video that shows the exercise being done in real time. You may find it helpful to watch a video before doing an exercise; the videos may also be a good refresher later. These GIS TV videos can be accessed from the Understanding GIS web site.

Other resources. The ArcGIS Desktop Help System, installed with the software and accessible online, provides comprehensive descriptions of software concepts and tools. Additional online resources such as blogs, forums, map galleries, videos, user communities, and access to technical support and training can be found at this website:

http://resources.arcgis.com

Disclaimer

The data used in this book is real. So are the efforts of the City of Los Angeles and several interest groups to improve the environmental quality of the Los Angeles River and its surroundings. The GIS project in this book, however, was developed entirely at Esri. For the sake of a story, we pretend that the project was sponsored by the Los Angeles City Council, and throughout the book there are references to what the "city council" wants or expects. In fact, neither the book nor the project have any affiliation with the city beyond permission to use its GIS data. Likewise, there is no affiliation with the Los Angeles River Revitalization Corporation or with any Los Angeles River advocacy organization.

Acknowledgments

This book would not have been possible without the cooperation of the City of Los Angeles, Department of Public Works, Bureau of Engineering. Special thanks to Randy Price and Ann-Kristin Karling of the Bureau's Mapping Division, and to City Engineer Gary Lee Moore, for giving us access to the city's data. We thank the City of Los Angeles Department of Recreation & Parks for providing its parks data. We note that land parcels and attributes are maintained by the Los Angeles County Assessor's Office.

The idea for the GIS project in this book was inspired by the Los Angeles River Revitalization Master Plan (available at http://www.lariver.org). Images from the master plan are used in Lesson 1 by permission.

Many people reviewed the book in whole or in part, tested exercises, and gave advice or help. Thanks to Mamata Akella, Richard Alden, Matthew Baker, Janel Day, Drew Flater, Charlie Frye, Koushik Hajra, Dale Honeycutt, Melita Kennedy, Deane Kensok, Natalie Matthews, Andy Mitchell, Eileen Napoleon, Brian Parr, Brandy Perkins, Tom Proctor, Veronica Rojas, Ryan Theis, Lindsay Thomas, Vin Thomas, and Molly Zurn.

Thanks to John Berry for participating in holistic testing of the book, and for testing and reviewing both the 1st and 2nd editions.

Thanks to Joyce Frye for copyediting the manuscript.

Thanks to Veronica Rojas for producing the GIS TV videos.

Thanks to Catherine Ortiz for management support.

Thanks to Peter Adams, David Boyles, Donna Celso, Kathleen Morgan, Brian Harris, Jennifer Hasselbeck, Jay Loteria, and Claudia Naber at Esri Press and to Cliff Crabbe for patiently shepherding the book through production.

And, most of all, a huge thanks to Clint Brown for supporting this project through thick and thin.

Technical requirements

For this book, your computer needs the following components:

- 2.2 GHz CPU with an Intel Core Duo, Pentium 4, or Xeon processor, 2 GB of RAM and at least 2.4 GB free disk space with up to 50 MB of disk space in the Windows System directory.
- Microsoft Windows XP, Vista, Windows 7, and Windows 8 operating system (with the most recent service packs).
- Microsoft .NET 3.5 SP1 must be installed prior to installing ArcGIS 10.1 for Desktop.
- ArcGIS 10.1 for Desktop software, Advanced license level (provided).
- Additional 800 MB free disk space for installing the exercise data (exercise data provided).
- ArcGIS for Desktop requires an installation of Microsoft Internet Explorer Version 7.0 or 8.0.
- An Internet connection.

Installing the free 180-day software trial of ArcGIS for Desktop software

Completion of the exercises in this book requires ArcGIS 10.1 for Desktop software (or newer version) at the Advanced license level. If you don't have the software, you can get a 180-day trial version at www.esri.com/UnderstandingGISforArcGIS10-1. Follow the on-screen instructions to complete the software installation. You will be prompted to enter the EVA code printed inside the back cover of this book. The trial is a fully functional, but nonrenewable, 180-day license.

Required software version

This book requires ArcGIS 10.1 for Desktop or later software with an Advanced license. Earlier software versions may not be fully compatible with exercise data and do not operate as described in the exercises.

Installing the project data

Follow the steps below to install the exercise data.

1) Put the data DVD in your computer's DVD drive. A splash screen will appear.

2) Read the welcome, then click the Install Exercise Data link. This launches the InstallShield Wizard.

3) Click Next. Read and accept the license agreement terms and click Next.

4) Accept the default installation folder or click Browse and navigate to the drive or folder location where you want to install the data.

5) Click Next. The installation will take a few moments.

6) Click Finish. The exercise data is installed on your computer in a folder called C:\UGIS (or \UGIS in the folder where you installed the exercise data).

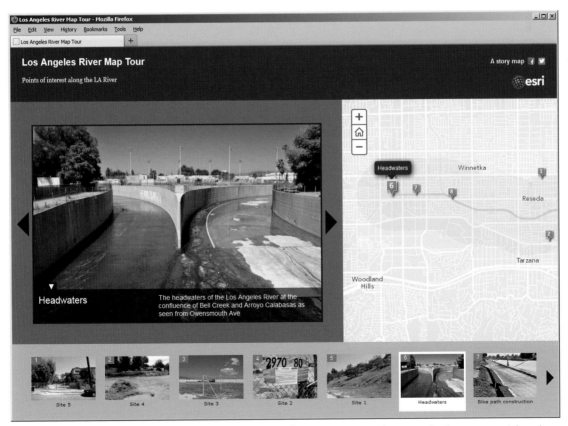

The Los Angeles River Map Tour is an Esri story map featuring recent photographs from potential park sites selected in the book, as well as other points of interest along the study corridor. The tour is located at ugis.esri.com/LA_River_Tour.

1 Frame the problem and explore the study area

THE VOLATILE LOS ANGELES RIVER

is the reason America's second-largest city was founded in its present southern California location by Spaniards in 1781. (The area was originally settled by the Gabrielino Native American Tribe thousands of years earlier.) Its water was tapped for drinking and irrigation, and a new city spread out from the river across the coastal plain. By the turn of the 20th century, the river was surrounded by a thriving urban center. Every few decades, raging floods would crest the banks at various points, submerging entire neighborhoods. After the historic floods of 1938 that claimed more than 100 lives and washed out bridges from Tujunga Wash to San Pedro, city leaders had seen enough. By 1941 the U.S. Army Corps of Engineers began to straighten, deepen, and reinforce the once wild waterway. Much of its length was eventually lined in concrete, and the river was more or less tamed.

Today, the City of Angels—home to nearly 4 million people—is a vibrant world center of business and culture. Running straight through the heart of the city, the Los Angeles River now serves ably as a flood-control channel. Sadly, this once-bucolic waterway that was so instrumental in the formation of the city became known as something ugly and marginal. Mile after mile of angled concrete appealed only to graffiti artists and filmmakers, and save for the occasional televised rescue of some hapless Angeleno swept away by a winter storm-fed torrent, the river remained a part of the city ignored by most. The negative perception has stuck with the neglected river for decades.

But in recent years, as the city has densified and much of southern California's wildlands have been appropriated for development, new attention has focused on the river corridor and the scattered pockets of open space that line its length. While it must always serve its important flood control function, the river and adjacent lands are increasingly recognized as under-utilized, providing opportunities for regreening and psychic restoration for human beings in an overbuilt city. Adventuresome and resourceful citizens have discovered peaceful pockets of sanctuary along the river and made these places their own. A vital and concerned activist community has raised awareness of the river and pushed for its beautification and redevelopment.

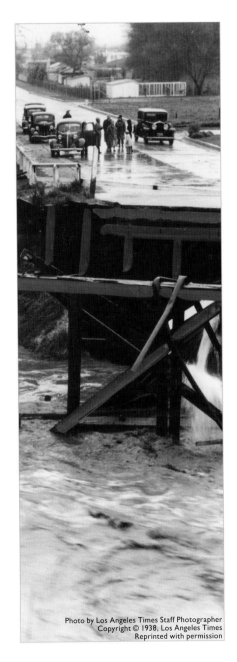

Photo by Los Angeles Times Staff Photographer
Copyright © 1938, Los Angeles Times
Reprinted with permission

In 2005, the city launched a major public works project focused on the human dimensions of the river. A landmark study, the Los Angeles River Revitalization Master Plan, demonstrated the significant potential to improve quality of life for citizens living near the river corridor through wise redevelopment. Mayor Antonio Villaraigosa said at the time: "We have an opportunity to create pocket parks and landscaped walkways...to create places where children can play and adults can stroll."

According to Villaraigosa, "The Plan provides a 25- to 50-year blueprint for transforming the City's 32-mile stretch of the river into an 'emerald necklace' of parks, walkways, and bike paths, as well as providing better connections to the neighboring communities, protecting wildlife, promoting the health of the river, and leveraging economic reinvestment."

While the 2005 master plan identified some of the most obvious areas for large-scale regional redevelopment along the river, it stopped short of identifying smaller (and more affordable) neighborhood projects; that work would require a more involved study. With thousands of land parcels strung out along the river, identifying the best places for park development is like looking for that proverbial needle in a haystack. Many factors come into play, among them current land use, demographics, and accessibility.

In the years since the plan's completion, the city has created a web site that encourages people to learn about (and participate in) the latest developments related to its landmark resource. The web site at www.lariver.org contains links to many resources about the river and its watershed, including scientific studies and recreational opportunities. If city leaders can find the resources and a motivated citizenry keeps up the pressure, there will be a renaissance transforming growing stretches along the river that become real versions of these artists' renderings.

All images on this page courtesy of The City of Los Angeles, LA River Revitalization Master Plan

Here's where you pick up the thread in this book. You'll use the city's real need for river redevelopment as a launching point for a park siting analysis using a geographic information system (GIS). A GIS is ideal for this type of decision-making because it allows you to analyze large amounts of data in a spatial context. In this book, you'll spend a lot of time with Esri's ArcGIS® for Desktop software, and by the end you'll have completed a project from start to finish. Along the way, you'll gain an excellent grasp of what a GIS can do.

You'll be assuming the role of a GIS analyst for the City of Los Angeles. So what exactly does a GIS analyst do and how is that job different from other jobs that also use GIS software? The table below defines some of the various roles that a typical GIS operation might establish to accomplish their work.

GIS Roles	
Editors	People who create, update, and correct spatial data and its attributes (statistical and descriptive information).
Cartographers	People who make and publish maps and solve information design problems.
Analysts	People who query and process geographic data to solve analytical problems.
Programmers	People who implement custom GIS functionality by developing scripts and applications for specific procedures.
Managers	People who oversee staffing and equipment, database design, workflow, new technology, and data acquisition.

The central work in this book is analytical. Your main focus will be on using ArcGIS tools and methods to find the most park-suitable land within a study area, but there is preparatory work to do before the analysis proper, and there are results to interpret and present afterwards. This book has two goals. One is to present a comprehensive approach to geographic problem solving. We want to help you develop skills, habits, and ways of thinking that will be useful in projects other than this one. The second goal is to teach you how to use ArcGIS 10.1 for Desktop software. These goals are mostly complementary. ArcGIS is a big system, however, and it wouldn't be realistic to try to cover all it can do in a single book. Our principle has been to teach the software in the service of the project and not otherwise. You'll delve into many aspects of ArcGIS for Desktop—editing, modeling, and cartography among them—but there are other aspects we won't use, or will only touch lightly, because they aren't strictly relevant to our needs. We might say (with apologies to Waldo Tobler) that everything in a GIS is related to everything else, but some things are more closely related to analysis than others.

Frame the problem

The first step in the geographic approach to problem-solving is to frame the problem. What that means, first of all, is coming up with a short statement of what it is you want to accomplish. We want to find a suitable site for a park near the Los Angeles River.

Once you have the statement, you can begin to tease out its ambiguities. What factors make a site "suitable"? Fortunately for us, the city council has already established a concise and fairly specific set of guidelines. They want a park to be developed:

- On a vacant parcel of land at least one acre in size
- Within the Los Angeles city limits
- As close as possible to the Los Angeles River
- Not in the vicinity of an existing park
- In a densely populated neighborhood with lots of children
- In a lower-income neighborhood
- Where as many people as possible can be served

This limits the scope of our inquiry, but it's far from a complete breakdown of the problem. Some of the guidelines are specific, but others are vague. Familiar concepts are sometimes the hardest to pin down. For example, what income level should count as "lower income"? How are the boundaries of a "neighborhood" established? We can't solve the big problem until we've solved the little problems buried inside it. Usually, however, it's not possible to address (or even foresee) all the little problems ahead of time.

Data exploration influences the framing of the problem. Do you have income data on hand? If so, is it for individuals or households? Is it average, median, or total? To the extent that the questions themselves are indefinite (what is lower income? how should it be measured?), the data you have available will help shape the answer.

Analysis also influences the framing of the problem. Given that we want a one-acre tract of vacant land, what do we do about adjacent half-acre lots? Is there a tool to combine them? If so, does it have undesired side effects, such as loss of information? Our data-processing capabilities (and our knowledge of them) may determine how we define a "one-acre parcel."

Even the results of an analysis influence the framing of the problem. Suppose, after having carefully defined the guidelines, we run the analysis and don't find any suitable sites. Do we report to the city council that there's just no room for a park anywhere? More likely, we change some of our definitions and run the analysis again.

Framing the problem, therefore, is an ongoing process, one that will occupy us through much of the book.

Lesson One roadmap

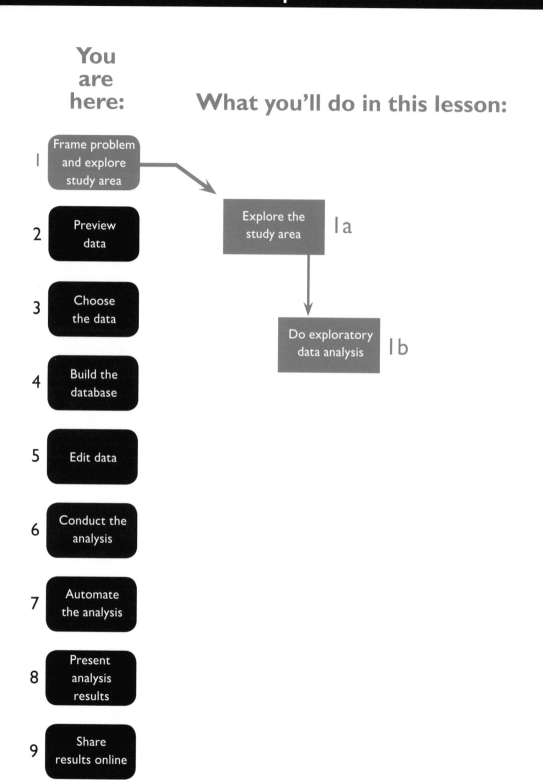

You are here:

What you'll do in this lesson:

1   Frame problem and explore study area

2   Preview data

3   Choose the data

4   Build the database

5   Edit data

6   Conduct the analysis

7   Automate the analysis

8   Present analysis results

9   Share results online

Explore the study area   1a

Do exploratory data analysis   1b

Exercise 1a: Explore the study area

In this exercise, we'll get to know the Los Angeles River and its surrounding area with maps and data. At the same time, we'll learn the basics of working with ArcMap: how to navigate a map, how to add and symbolize data, and how to get information about map features.

1) **Start ArcMap.** We'll open the ArcMap application.

Ⓐ Click the Windows Start button and choose All Programs → ArcGIS → ArcMap 10.1.

The application opens with the ArcMap - Getting Started dialog box in the foreground (Figure 1-1).

Ⓑ In the left-hand window, if necessary, click My Templates.

Other templates are listed here.

A template is a predefined layout, or design, for a map. It may include graphic elements and data. The default template is a blank map.

Figure 1-1

Ⓒ Click the Blank Map icon to select it and click OK.

A blank map document opens. A map document is a file that contains one or more maps, layouts, and associated layers, tables, charts, and reports.

2) **Add a basemap layer.** A basemap is a layer of reference geography that forms a backdrop to your map.

Ⓐ On the Standard toolbar at the top of the application window, locate the Add Data button ⊕▾ .

Ⓑ Click the small drop-down arrow next to the button.

▷ If you click the icon instead of the arrow, an Add Data dialog box will open. Close this dialog box and try again.

Ⓒ On the drop-down menu, choose Add Basemap.

❶ In the Add Basemap window, click *Streets* to select it (Figure 1-2), then click Add.

Figure 1-2

Click the *Streets* basemap.

You'll probably be prompted to enable hardware acceleration. (It depends on your computer configuration.)

❷ If you're prompted to enable acceleration, click Yes.

This book makes frequent use of online map services; specifically the *Streets* and *Imagery* basemaps. One side effect is that there can be complications due to servers, web and network connections, and the processing power of your computer. Therefore, operations in some cases may be slower than you anticipate. Be reasonably patient as you work, and check the Understanding GIS web site (see page xii) for updated information if performance issues persist.

The *World_Street_Map* basemap layer is added to your map document (Figure 1-3). It has an entry in the Table of Contents window on the left. At the top of the application is the main menu. Underneath it are two toolbars. The Standard toolbar has generic functions (such as New, Save, Print, Copy, Paste) and buttons to open other windows. The Tools toolbar has navigation tools (such as Zoom In, Zoom Out, and Pan) and other common tools for doing things with maps. There are lots of other toolbars besides these, and we'll use a couple of them in this lesson.

Figure 1-3

On the right side of the application are two tabs named Catalog and Search. These tabs are hidden windows. We'll use both windows a lot in the book, although we won't get to the Search window for a few more lessons.

Toolbars and windows can be moved around and resized. Figure 1-3 reflects the default arrangement.

3) Zoom in to Southern California. We'll zoom in to our area of interest.

Ⓐ On the Tools toolbar, click the Zoom In tool 🔍. In the map window, click and drag to draw a box around Southern California, as shown in Figure 1-4.

▷ Your box doesn't have to match exactly.

Ⓑ Use the Zoom In tool again to get closer. When you see city names and major roads, use the Pan tool 🖑 to center the view on the Los Angeles area.

Ⓒ Keep zooming in (try the Fixed Zoom In button ⛶, too) until you can easily distinguish cities, freeways, and landmarks such as airports, as in Figure 1-5.

▷ If you zoom in further than you want, use the Zoom Out 🔍 or Fixed Zoom Out ⛶ tools to go back.

You probably noticed that no streets were visible at the global scale, and that as you kept zooming in, more and more detail appeared. This is because the basemap is a multiscale map: really a set of maps which turn themselves on and off to display features and symbology that are appropriate to your map scale.

Figure 1-5 shows greater Los Angeles, an area that includes hundreds of incorporated communities and nearby cities.

Figure 1-4

Draw a box around the area you want to zoom to.

Figure 1-5

Basemap layers

Basemap layers show reference geography such as street maps, imagery, topography, and physical relief. The ones available from the Add Basemap window in ArcMap® are remotely hosted map services that you can navigate, view, and use as backdrops to other data. Basemaps are stored at multiple scales, so that as you zoom in or out you see different amounts of detail. As you navigate a basemap, the various pieces of it (called *tiles*) that compose your current view are stored locally on your computer in a so-called display cache. When you zoom or pan to a new area, the map may be a little slow to draw, but any place you return to will redraw quickly because the data comes from your cache, not from the remote server.

Figure 1-6

4) **Add a layer of project data.** On top of the basemap, we'll add a layer from the data that has been put together for this project and is stored on your computer. In Lesson 2, we'll talk more about where this data comes from and how to acquire data of your own.

Ⓐ On the Standard toolbar, click the Add Data button ✛. This time, click the icon itself, not the drop-down arrow.

The Add Data dialog box opens to a default "home" location in your Documents\ArcGIS folder. We don't have any project data here. To navigate to our data folder, we have to make a folder connection.

Ⓑ In the row of buttons at the top of the Add Data dialog box, click the Connect To Folder button ⊞.

Ⓒ In the Connect To Folder dialog box, navigate to the folder where you installed the data—probably C:\UGIS— (Figure 1-6) and click OK.

Ⓓ In the Add Data dialog box, navigate as follows:
  • Double-click the ParkSite folder to open it.
  • Double-click the Source Data folder.
  • Double-click ESRI.gdb.
  • Double-click Boundary.

We'll discuss GIS data formats in the topic *Representing the real world as data* in Lesson 2. For now, we just want to dig down to our data.

Ⓔ In the Add Data dialog box, click *City_ply* to select it (Figure 1-7) and click Add.
  ▷ Or just double-click *City_ply* from the list.

Figure 1-7

A layer of city boundaries is added to the map (Figure 1-8). Each city in the layer is called a feature. These features are polygons, which are one of the three basic shapes with which geographic objects are represented in a GIS. (The others are lines and points.)

The color of the layer is assigned at random, so yours may be different.

Figure 1-8

In the Table of Contents, the order of entries matches the drawing order of layers in the map. *City_ply* is listed above *Basemap,* and on the map, the cities cover the basemap. You can control a layer's visibility with its check box in the Table of Contents.

F In the Table of Contents, click the check box next to *City_ply*.

The layer turns off.

G Click its check box again to turn the layer back on.

5) Set layer properties. Every layer has properties you can set and change. For example, you just changed the visibility property of the *City_ply* layer.

A In the Table of Contents, right-click the color patch underneath the *City_ply* layer name.

A color palette opens (Figure 1-9). Moving the mouse pointer over any color square shows its name as a tool tip.

B On the color palette, click any color you like to change the layer color.

On the map, the cities redraw in the color you chose.

Right-click here

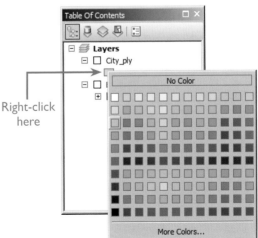

Figure 1-9

C In the Table of Contents, right-click the *City_ply* layer name to open its context menu. At the bottom of the menu, choose Properties.

The Layer Properties dialog box opens. This is where you access the full set of properties for a layer.

D If necessary, click the General tab (Figure 1-10).

Each tab along the top has different layer properties.

Figure 1-10

The layer's name, *City_ply,* is one of its properties. This name is cryptic (it stands for "city polygons") and unattractive, so let's change it.

E In the Layer Name box, delete the name and type **Cities** instead. Click Apply.

The name is updated in the Table of Contents.

F In the Layer Properties dialog box, click the Source tab.

This tab shows technical information about the layer, including the path to the data on your computer (Figure 1-11).

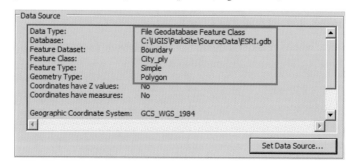

Figure 1-11

Renaming the layer in the map document doesn't change the name associated with the source data (which is still *City_ply*). A layer is a representation or rendering of the data, not the data itself. You can make any changes you want to a layer's properties without affecting the data on which the layer is based.

Ⓖ Click OK on the Layer Properties dialog box to close it.

6) Get information about cities. Let's see what we can find out about the cities on the map.

Ⓐ On the Tools toolbar, click the Identify tool 🛈.

Ⓑ Click any city polygon on the map.

The city flashes green and an Identify window opens (Figure 1-12). In the top window, you see the name of the city you identified. In the bottom window, you see its attributes, or the information that this layer stores about cities. Some of them aren't obviously meaningful, but others are. If POP2010 is population for the year 2010, and POP10_SQMI is population per square mile for the same year, then you know that Los Angeles (if that's the city you identified) had 3,792,621 inhabitants at the time of the 2010 census and its population density was 8,018 people per square mile.

Ⓒ Move the Identify window away from the map, if necessary. Click another city on the map.

The Identify window updates with information about the new city. All the cities have the same set of attributes; it's the values of the attributes that change.

Ⓓ Identify a few more cities, then click somewhere on the basemap where there isn't a city.

The Identify window is empty and a message at the bottom says, "No identified features."

Ⓔ Close the Identify window.

Ⓕ In the Table of Contents, right-click *Cities* and choose Open Attribute Table (Figure 1-13).

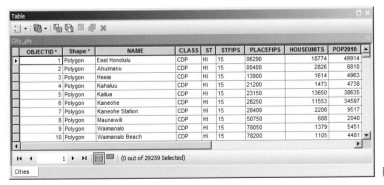

Figure 1-12

This example shows Los Angeles, but it doesn't matter which city you identify.

Figure 1-13

G Scroll across the table and look at the field names (the gray column headings).

This is a different presentation of the same information you saw when you identified cities.

H Scroll back to the beginning of the table, then scroll a little way down through the records (the table rows).

There are a lot of records in this table: in fact, 29,259 of them, as you can see at the bottom. Each corresponds to a unique feature—that is, a unique city—on the map. So there must be a lot of cities we're not seeing in the current view.

I Leave the table open, but move it out of the way of the map.

▷ You can dock it by dropping it on any blue arrow, or you can move it outside the ArcMap application window.

J In the Table of Contents, right-click the *Cities* layer. On the context menu, choose Zoom To Layer.

The map zooms out to the geographic extent of the layer: the entire United States. You can't distinguish individual cities at this scale.

K On the Tools toolbar, click the Go Back To Previous Extent button ⬅.

7) Select the record for Los Angeles. When you select a record in an attribute table, the corresponding feature is selected on the map. (Likewise, when you select a feature on the map, its record is selected in the table.) Selections are marked with a blue highlight.

A Scroll up to the top of the table. Make sure the table is wide enough that you can see the POP2010 field.

▷ If necessary, widen the table by dragging its edge.

B Right-click the POP2010 field name. On the context menu, choose Sort Descending.

The records are sorted in the order of their populations, from largest to smallest. Los Angeles is now the second record in the table, right after New York.

C At the left edge of the table, click the small gray box next to the Los Angeles record to select it (Figure 1-14).

2. Click the box to select the record.

1. Right-click and choose Sort Descending.

Figure 1-14

On the map, the city of Los Angeles is highlighted in blue. You may be able to see the whole city already, but let's make sure.

D Close the attribute table.

E In the Table of Contents, right-click *Cities* and choose Selection → Zoom To Selected Features.

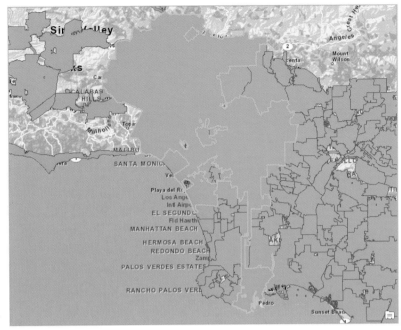

Figure 1-15

The map zooms in close on the selected feature and centers it in the view (Figure 1-15). The city's odd shape is attributable to years of piecemeal expansion and incorporation. It has internal "holes" where it surrounds other cities, such as Beverly Hills, or unincorporated areas. It also has a long, narrow southern corridor that connects it to its harbor at the Port of Los Angeles.

F On the Tools toolbar, click the Clear Selected Features button 🔽 to unselect the feature.

8) **Filter the display of cities with a definition query.** One of our project requirements is that the new park be inside the Los Angeles city limits. Therefore, we're more interested in Los Angeles than in other cities. A layer property called a definition query lets us show only those features in a layer that interest us.

A In the Table of Contents, right-click the *Cities* layer and choose Properties.

B In the Layer Properties dialog box, click the Definition Query tab.

C Click Query Builder to open the Query Builder dialog box (Figure 1-16).

Figure 1-16

You build a query on an attribute table by specifying a field and setting a logical or arithmetic condition that values in that field have to satisfy. In our case, we want to find records with the value "Los Angeles" in the NAME field.

D In the list of field names at the top of the Query Builder dialog box, double-click "NAME."

The field is added to the expression box at the bottom of the Query Builder.

E Click the Equals button [=] .

F Click Get Unique Values.

The middle box in the Query Builder is populated with a list of all the city names in the NAME field. Since there are so many, it would be inconvenient to scroll to the one we want.

G Click in the Go To box and start typing **Los Angeles**.

By the time you get to "Los An" the value "Los Angeles" will be highlighted in gray.

H Double-click Los Angeles.

I Compare your expression to Figure 1-17. If it doesn't match, click Clear and rebuild it.

Figure 1-17

J Click Verify and click OK on the Verifying Expression message box.

▷ If you get an error message, click OK. Clear and rebuild the expression.

K Click OK on the Query Builder dialog box.

The expression appears on the Definition Query tab (Figure 1-18).

❶ Click OK on the Layer Properties dialog box.

On the map, only the city of Los Angeles is shown. The other cities are hidden by the definition query (Figure 1-19).

Figure 1-18

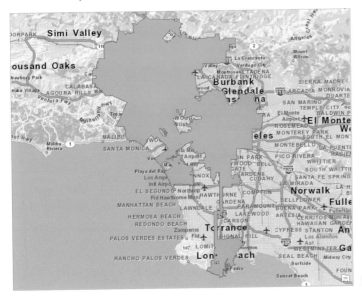

Figure 1-19

❿ In the Table of Contents, right-click the *Cities* layer and choose Open Attribute Table.

The table shows just the record for Los Angeles.

❶ Close the attribute table.

The other city features haven't been deleted. Clearing the definition query would display them again. Layer properties affect the display of data in a map, not the essential properties of the data itself: the number of features, their shapes, locations, and attributes.

Layers and datasets

A layer points or refers to a dataset stored somewhere on disk (as specified on the Source tab of the Layer Properties dialog box). A layer is *not* a physical copy of the data. A layer is a representation or rendering of the data. The dataset stores the shapes, locations, and attributes of features; the layer stores display properties, including what the layer is named, which of its features are shown or selected, how those features are symbolized, and whether they are labeled. Changes to layer properties do not affect the dataset that the layer refers to. You can make as many layers as you want from the same dataset and give them different properties. These layers can coexist in the same map document or in many different map documents.

9) **Add a layer of rivers.** Let's add a layer of local rivers to the map and see where the Los Angeles River fits into the picture.

Ⓐ On the Standard toolbar, click the Add Data button ⊞. (Click the icon itself, not the drop-down arrow.)

The Add Data dialog box opens to your last location. There's no river data here.

Ⓑ In the row of buttons at the top of the Add Data dialog box, click the Up One Level button 🔼 (Figure 1-20).

Figure 1-20

Ⓒ Double-click Hydro.

Ⓓ Click *River* to select it and click Add.

A layer of rivers is added to the map and placed at the top of the Table of Contents.

Ⓔ On the Tools toolbar, click the Identify tool ⓘ, if necessary.

Ⓕ Click on any river to identify it.

You see the name of the river (many of them don't have names) and its other attributes (Figure 1-21).

By default, the Identify tool identifies features from the topmost layer in the Table of Contents. If you miss a river, you'll either identify the City of Los Angeles or nothing. That's fine—leave the Identify window open and click again on a river.

Ⓖ Click on a few more rivers to identify them. Try to identify a segment of the Los Angeles River.

The river runs west to east across the city, turns south near the city's eastern edge, and follows a freeway to San Pedro Bay.

Ⓗ Close the Identify window.

10) **Make a definition query on the LA River.** Just as we're mainly interested in one city, we're mainly interested in one river. We'll make another definition query to show just the Los Angeles River.

Ⓐ In the Table of Contents, right-click the *River* layer and choose Properties.

▷ A shortcut is to double-click the layer name in the Table of Contents.

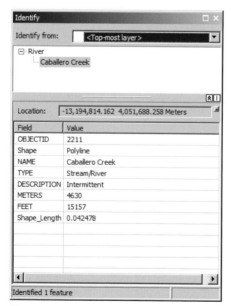

Figure 1-21

ⓑ Click the Definition Query tab, if necessary.

ⓒ Click Query Builder to open the Query Builder dialog box.

ⓓ In the list of field names at the top of the Query Builder dialog box, double-click "NAME."

ⓔ Click the Equals button ⌐ = ⌐.

ⓕ Click Get Unique Values.

ⓖ Click in the Go To box and start typing **Los Angeles River**.

The value will be highlighted in gray before you finish typing.

ⓗ Double-click Los Angeles River.

ⓘ Compare your expression to Figure 1-22. If it doesn't match, click Clear and rebuild it.

ⓙ Optionally, verify the expression.

ⓚ Click OK on the Query Builder dialog box.

ⓛ On the Layer Properties dialog box, click the Symbology tab.

ⓜ On the Symbology tab, click the button displaying the blue line symbol to open the Symbol Selector.

ⓝ Change the line width to 3 (Figure 1-23) and click OK on the Symbol Selector.

Figure 1-22

Figure 1-23

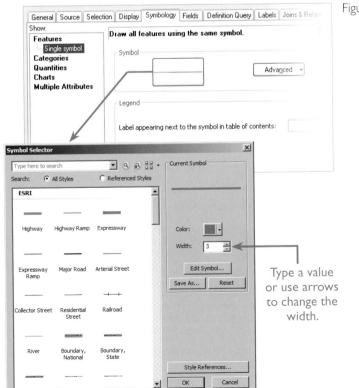

O On the Layer Properties dialog box, click the General tab.

P In the Layer Name box, delete the layer name *River* and type **Los Angeles River** (Figure 1-24).

▷ You can also rename a layer directly in the Table of Contents by clicking the name once to highlight it and clicking it again. (Be careful not to double-click or you'll open the layer properties.)

Figure 1-24

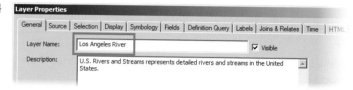

Q Click OK on the Layer Properties dialog box.

On the map, the river is displayed with its new symbology (Figure 1-25). The Table of Contents reflects the new layer name.

11) Select Los Angeles River features with a query. On the map, the river looks like a single feature (just like the city), but it's not.

A In the Table of Contents, right-click the *Los Angeles River* layer and choose Open Attribute Table.

At the bottom of the table, you see that 0 of 17 records are selected. That means that, in this particular layer, the Los Angeles River is composed of 17 features. Why would that be?

B Scroll down through the table.

Figure 1-25

All the records have the same NAME value. Most have the same TYPE, but one is an artificial path. There are a few DESCRIPTION values. The need to maintain different attribute values for different parts of a geographic object is a common reason that data—especially data representing linear features such as streets and rivers—is constructed this way (Figure 1-26). We'll come back to this point in the next lesson.

Figure 1-26

Ⓒ In the row of tools at the top of the table window, click the Select By Attributes button .

Having noticed "perennial" and "intermittent" values in the DESCRIPTION field, we might want to know which parts of the river are which. We can find out with an attribute query. An attribute query is like a definition query in that both single out features in a layer on the basis of attribute values. The difference is that an attribute query highlights (selects) features that satisfy the expression rather than hiding features that don't.

Ⓓ At the top of the dialog box, make sure the Method drop-down list is set to "Create a New Selection."

Ⓔ In the list of field names, double-click "DESCRIPTION" to add it to the expression box.

Ⓕ Click the Equals button [=].

Ⓖ Click Get Unique Values.

Ⓗ In the list of unique values, double-click 'Perennial.'

Ⓘ Confirm that your expression matches the one in Figure 1-27 and click Apply.

Figure 1-27

Twelve records are selected in the table. The corresponding features are selected on the map, showing you that the river is perennial for most of its length (Figure 1-28).

It's not among our guidelines to locate the new park along the river's perennial stretch, but it's interesting that fairly simple data exploration may introduce new ways of thinking about your problem.

Ⓙ Close the Select By Attributes dialog box.

Ⓚ At the top of the table window, click the Clear Selection button 🔲 .

Clearing the record selection clears the feature selection on the map. (The reverse is true as well.)

Ⓛ Close the attribute table.

Selecting features

You make selections in order to work with a subset of features in a layer. You can use a selection to zoom the map in to a specific area, to make a new layer from the selected subset, to get statistical information about the subset, or for many other things.

Selections are also used in queries. Whether you do an attribute query (to find records with a certain attribute value) or a spatial query (to find features with a certain spatial relationship to other features), the records and features that satisfy the query are returned as a selection on the layer.

Figure 1-28

12) Save the map document. This is a good time to save the map document. Save your maps often as you do these exercises—don't wait to be reminded.

Ⓐ On the Standard toolbar click the Save button 💾.

In the Save As dialog box, ArcMap should default to your My Documents\ArcGIS folder. In this book, we'll save most of our maps to the ParkSite\MapsAndMore folder.

Ⓑ In the panel of icons on the left side of the dialog box, click Computer.

Ⓒ Navigate to C:\UGIS\ParkSite\MapsAndMore.

Ⓓ When you're inside the MapsAndMore folder, replace the default file name with **Lesson1** (Figure 1-29) and click Save.

Click here to navigate

Figure 1-29

13) Add a basemap of imagery. Imagery provides a detailed, photorealistic view of the ground, and we'll rely on it to explore the LA River in more detail. Imagery also has other important uses, such as providing a background against which to edit features (Lesson 5), and "ground-truthing" analysis results (Lesson 6).

Ⓐ On the Standard toolbar, click the drop-down arrow next to the Add Data button and choose Add Basemap.

Ⓑ In the Add Basemap window, click *Imagery with Labels* to select it (Figure 1-30) and click Add.

The selection and arrangement of basemaps is subject to change, so your screen may look different.

Figure 1-30

The *World_Imagery* basemap layer is added to the bottom of the Table of Contents. You don't see it on the map because it's covered by the basemap of streets. A layer called *Reference* has simultaneously been added to the top of the Table of Contents. This layer consists of locations, boundaries, and labels and should be visible on the map. (You should see yellow place-names covering the city of Los Angeles feature, for instance.).

> ▷ If you don't see these labels, click the Refresh button ⟳ in the lower left corner of the map window.

We don't need two basemaps in the map, so we'll remove the basemap of streets.

C In the Table of Contents, right-click the name *Basemap* above *World_Street_Map* (Figure 1-31) and choose Remove.

The basemap is removed and the imagery is now visible. Your Table of Contents and map should look like Figure 1-32.

Figure 1-31

Right-click here.

Here the Table of Contents is shown floating free. By default it's docked in the ArcMap application window, which is probably where it is on your screen.

Figure 1-32

Create Bookmark

Bookmark Name: City of Los Angeles

OK   Cancel

Figure 1-33

14) Create a bookmark. Certain views in a map are useful for orientation or reference. You can bookmark a view to make it easy to return to.

Ⓐ From the main menu, choose Bookmarks → Create Bookmark.

Ⓑ In the Create Bookmark dialog box, replace the default name with **City of Los Angeles** (Figure 1-33). Click OK.

Ⓒ Use the Zoom In tool 🔍 to zoom in on the river's mouth in the harbor at Long Beach.

Ⓓ On the Standard toolbar, click the Map Scale drop-down arrow and choose 1:10,000 (Figure 1-34).

At this scale, one unit of measure on the map is equivalent to ten thousand of the same units on the ground. Loosely, a thing on the map is ten thousand times smaller than its actual size.

Figure 1-34

Ⓔ Use the zoom and pan tools to explore the harbor.

The imagery is very high resolution and you'll see more detail as you keep zooming in. Eventually, you'll reach a limit.

Ⓕ When you're ready, from the main menu, choose Bookmarks → City of Los Angeles.

The map view returns to the bookmarked extent.

15) Change the symbology for Los Angeles. The boundary of Los Angeles is filled in with a solid color, so the imagery underneath is still covered up. Pretty soon we're going to zoom in and follow the river's course through the city. It will serve that purpose to resymbolize the city so that we just see its outline.

A Open the layer properties for the *Cities* layer. Click the Symbology tab, if necessary.

B On the Symbology tab, click the button displaying the current color symbol to open the Symbol Selector.

C In the Symbol Selector, click the Fill Color button to open the color palette.

▷ You can click either the color square itself or the drop-down arrow.

D At the top of the color palette, click No Color.

This will make the symbol transparent.

E Change the Outline Width value to 2.

F Click the Outline Color button. On the color palette, click Autunite Yellow (Figure 1-35).

Figure 1-35

Figure 1-36

G Click OK on the Symbol Selector. Click Apply on the Layer Properties dialog box and leave it open.

The new symbology is displayed on the map (Figure 1-36).

H Click the General tab and rename the layer **Los Angeles**.

I Click OK on the Layer Properties dialog box.

16) Save the layer as a layer file. It wasn't hard to make this particular symbol, but often it does take time and effort to create good symbology. Having done so, you may want to reuse that symbology. In this book, for instance, we'll draw the outline of Los Angeles in other map documents in coming lessons.

You can save layer properties to a file called a layer file, which has the file extension *.lyr*. A layer file is not a copy of the data, but you add it to a map document in the same way that you add data. The layer file stores all the properties of a layer—name, symbology, definition query, and so on—including the path to the layer's source dataset. When you add a layer file to ArcMap, the layer draws with its properties already set.

A In the Table of Contents, right-click *Los Angeles* and choose Save As Layer File (near the bottom of the context menu).

The Save Layer dialog box opens to the ParkSite\MapsAndMore folder (Figure 1-37).

B Accept the default name *Los Angeles.lyr* and click Save.

Figure 1-37

When you saved the map document in the MapsAndMore folder, this became the map's "home" folder. Now other files associated with the map, such as layer files, will be saved to this location by default.

We'll remove the *Los Angeles* layer that's currently in the map, then add the layer file to see how it works.

C In the Table of Contents, right-click *Los Angeles* and choose Remove.

The layer disappears from the map and the Table of Contents.

D On the Standard toolbar, click the Add Data button.

E In the Add Data dialog box, click the Look in: drop-down arrow and click the ParkSite folder (Figure 1-38).

F In the dialog box, double-click the MapsAndMore folder.

G Click *Los Angeles.lyr* to select it (Figure 1-39) and click Add.

Figure 1-38

Figure 1-39

The layer is added to the map with its properties set.

H Open the layer properties for *Los Angeles*. In the Layer Properties dialog box, click the Source tab.

As before, the layer points to the *City_ply* feature class.

I Close the Layer Properties dialog box.

Anytime you add the *Los Angeles.lyr* file to a map document, the layer will draw as a yellow outline with a definition query on the city of Los Angeles. Once the layer has been added to the map, however, it's just the same as any other layer, and you can change its properties however you want. (Not that you want to change them.)

17) Follow the river. Let's start developing a sense of the study area by following the river's course through the city.

A Maximize the ArcMap window if you haven't done so already.

B Use the Zoom In tool to zoom in on the river's source in the community of Canoga Park in northwest Los Angeles.

The river officially starts where Bell Creek and the Arroyo Calabasas converge at Canoga Park High School.

C On the Standard toolbar, click the Map Scale drop-down arrow and choose 1:24,000.

By default, map scale is displayed in the form of a representative fraction, such as 1:24,000. This means, for any unit of measure, that one unit of distance on the map is equivalent to 24,000 units in the real world. You can set the map to any scale you want by typing a number into the scale box and pressing Enter. You can also enter a verbal expression such as "1 inch = 1 mile" and it will be converted to a representative fraction. To change the way scale is displayed, or to change the list of predefined scales, click <Customize This List> at the bottom of the Map Scale drop-down list.

Figure 1-40

Sepulveda Dam Recreation Area

D Click the Pan tool and slowly pan eastward along the river.

Densely populated residential neighborhoods line both sides of the river until you get to the Sepulveda Dam Recreation Area (Figure 1-40), a large recreational area with golf courses and a lake. The river bottom is natural here, becoming concrete again at the Sepulveda Dam in the southeastern corner of the basin.

East of the Sepulveda Basin, the river follows a freeway for a while, and is again surrounded by fairly dense residential and commercial areas.

E On the keyboard, press and hold the Q key.

Now you roam continuously across the display in whichever direction you point the mouse. To control your speed, make small brushing movements with the mouse either with or against the direction of movement. As you roam, the imagery should draw smoothly and continuously, although your experience may vary. Other layers, such as the *Los Angeles River* layer, suspend drawing and catch up when you stop.

F Release the Q key to stop roaming.

  ▷ You can also use the four arrow keys on the keyboard to roam.

G Continue to pan (or roam) along the river.

The river flows generally southeast for a while, then follows the northern edge of unincorporated Universal Studios (Figure 1-41). It continues east, then bends sharply south as it curves around Griffith Park (at 4,218 acres, one of the largest city parks in the United States). To the north of Griffith Park lies the city of Burbank; to the east is Glendale.

As it flows south, the river runs parallel to another major freeway. You'll see the Silver Lake Reservoir and then Elysian Park, where the Los Angeles Dodgers play major league baseball.

Figure 1-41

Universal Studios

H **Save bookmarks as place files.** Dodger Stadium is a landmark that we may want to return to.

I Center your view on Dodger Stadium, more or less as shown in Figure 1-42.

J From the main menu, choose Bookmarks → Create Bookmark.

K In the Create Bookmark dialog box, name the bookmark **Dodger Stadium** (Figure 1-43) and click OK.

Figure 1-42

Figure 1-43

Bookmarks are meant for use within a particular map document. The two we've created so far, however, might be useful in other maps. There's a tool for managing saved places so that any map document can access them: it's called My Places and we'll use it in Lesson 3. For now, we'll save our bookmarks as external files. When the time comes, we'll be able to load them into My Places.

L From the main menu, choose Bookmarks → Manage Bookmarks.

You should see two bookmarks in the Bookmarks Manager, with City of Los Angeles highlighted on top (Figure 1-44).

M Near the bottom of the Bookmarks Manager dialog box, click Save → Save All.

Figure 1-44

Figure 1-45

The Save Bookmarks dialog box opens to the MapsAndMore folder.

 In the File name box, type **Lesson 1 Places** (Figure 1-45) and click Save.

 Close the Bookmarks Manager.

18) Pan to the city limits. We'll follow the river until it crosses the LA city limits, which marks the boundary of our study area.

 Pan along the river as it runs south.

This last section of river passes through an industrial landscape and leaves the city at the Redondo Junction train yards.

 Save the map.

 If you're going on to the next exercise now, leave ArcMap and Lesson1.mxd open; otherwise, from the main menu, choose File → Exit.

Exercise 1b: Do exploratory data analysis

In this exercise, we'll add park data and census data (containing demographic and socioeconomic information) to our map. Our goal is to pay attention to patterns in the data and thereby build an intuitive sense of likely and unlikely locations for the park. This intuition should give us confidence that the analysis results we get in Lesson 6 are plausible. Conversely, if the results contradict our gut feeling, we may be alerted to possible mistakes in the analysis.

1) Start ArcMap. We'll start ArcMap, if necessary, and continue working with our map document from the last exercise.

 If ArcMap is open from the last exercise, go to Step 2.

 Otherwise, start ArcMap.

▷ You may want to pin ArcMap to your Start menu or create a desktop shortcut to it.

In the ArcMap - Getting Started dialog box, you should see Lesson1 under the heading "Recent."

⊙ Click Lesson1 to highlight it (Figure 1-46) and click Open.

Figure 1-46

The map opens to a view of where the LA River leaves Los Angeles and enters the industrial city of Vernon.

2) Add a layer of parks. One of our geographic constraints is that the new park not be located too close to existing parks. We know there are a couple of big parks along the river, but let's look at the whole situation.

Ⓐ On the right side of the ArcMap application window, click the Catalog tab to open the Catalog window (Figure 1-47).

▷ If you don't have this tab, on the Standard toolbar, click the Catalog button 🗐.

The Catalog window lets you view and manage geographic files on disk. The top entry in the window is your home folder, which presently contains your map document, the layer file you saved in the last exercise, and some graphs that are part of the starting data and which you'll use in Lesson 8. By default, the catalog recognizes only a limited number of file types to help you focus on geographic data.

The Catalog window has two states. When the pin icon in the upper right corner points sideways, the window closes if not in focus. When the pin icon points down, the window stays open. In this state, you can move and resize it. Leave the window in whichever state you prefer.

Ⓑ In the Catalog window, click the plus sign next to Folder Connections to expand it.

⊙ Expand C:\UGIS.

Click the Auto Hide button to toggle the window's state

Figure 1-47

If you don't see Los Angeles.lyr, right-click the Home folder and choose Refresh.

Figure 1-48

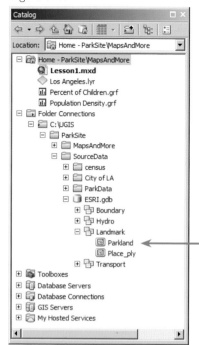

In figures, the Catalog window is usually shown floating free from ArcMap (as here).

This is the folder connection you created in the last exercise from the Add Data dialog box.

D Expand the following items:

- ParkSite
- SourceData
- ESRI.gdb
- Landmark

The Catalog window should look like Figure 1-48.

E In the Catalog window, click *Parkland* to select it.

F Drag and drop *Parkland* anywhere over the ArcMap window.

The layer is added to the Table of Contents and the map. Dragging and dropping data from the Catalog window is the same as adding it with the Add Data button—it's just faster.

3) **Symbolize the layer.** A shade of green is usually the right cartographic choice for parks. ArcMap may have chosen one by a stroke of luck, but probably not. We'll symbolize the layer after taking a look at its extent.

A In the Table of Contents, right-click the *Parkland* layer and choose Zoom To Layer (Figure 1-49).

Figure 1-49

B In the Table of Contents, turn *Parkland* off and on a few times by clicking its check box. Leave it checked on.

The data extends well beyond Los Angeles, and the basemap shows that those big polygons to the north correspond to mountains. They're probably national forests. Notice that the *Parkland* layer is also obscuring some of the labels on the map.

C In the Table of Contents, click the *Reference* layer to highlight it. Drag it to the top of the Table of Contents (its position is marked by a horizontal black line) and drop it above the *Parkland* layer.

The labels now draw on top of the park features.

D In the Table of Contents, click the color patch under the *Parkland* layer name to open the Symbol Selector (Figure 1-50).

  ▷ This is a shortcut to opening the Symbol Selector from the layer properties.

E In the Symbol Selector, click the Fill Color button to open the color palette. Click Apple Dust (Figure 1-51).

F Click the Outline Color button to open the color palette again. Click Moss Green (Figure 1-51).

G Click OK on the Symbol Selector.

The layer symbology is updated on the map.

4) **Identify features.** Let's find out what attributes this layer has.

A On the Tools toolbar, click the Identify tool 🛈 .

B Click on one of the big park features in the mountains to identify it.

Among the attributes are the park's name, its type, and its acreage (Figure 1-52).

Figure 1-50

Figure 1-51

Figure 1-52

C Leave the Identify window open. From the main menu, choose Bookmarks → City of Los Angeles.

At this scale, you can make out the larger parks within the city.

D Try to locate the parks mentioned on page 28: the Sepulveda Dam Recreation Area, Griffith Park, and Elysian Park. Click on each with the Identify tool 🛈 to see their attributes.

E When you're done, leave the Identify window open and zoom to the Dodger Stadium bookmark.

Figure 1-53

ArcMap places labels dynamically. The position of labels in a view depends on the map scale, the extent, the presence of other features, and more. The label placement in your map may be different from what's shown here.

❺ Identify some of the neighborhood-sized parks in the view.

❻ Close the Identify window.

Our list of park requirements doesn't have an upper size limit, but we can expect our candidates to be under ten acres. Our trip down the river in the last exercise didn't reveal any great big tracts of open land that weren't already parks.

5) Label parks. Some of the larger parks are already labeled in the *Reference* layer, but many of the smaller ones are not.

❶ In the Table of Contents, right-click *Parkland* and choose Label Features.

The parks are labeled with their names in simple black text (Figure 1-53). The information is being taken from the NAME attribute in the *Parkland* attribute table.

The default labels aren't ideal for this map. The black text disappears into the imagery wherever the label doesn't fit inside the park. In addition, the labels appear on one line no matter how long they are.

❷ Open the layer properties for the *Parkland* layer and click the Labels tab (Figure 1-54).

Figure 1-54

If the layer attribute table has a NAME field, ArcMap uses it by default.

❸ In the Text Symbol area, click the font size drop-down arrow and change the size from 8 to 9 (points).

❹ Click the color button. On the color palette, change the text color to Lemongrass (Figure 1-55).

Figure 1-55

Light green text will show up better than black, but to be legible it will still need an outline, or halo.

E Click the Symbol button to open the Symbol Selector.

F On the Symbol Selector, underneath the properties of the current symbol, click Edit Symbol.

G In the Editor dialog box, click the Mask tab.

H Click the Halo option and change the size to 1.

The default halo color is white, but we're going to use something darker.

I Click Symbol.

The Symbol Selector opens again.

J Click the Fill Color button and change the fill color (the halo color) to Moss Green (Figure 1-56).

K Click OK on the Symbol Selector.

Your Editor dialog box should look like Figure 1-57.

Figure 1-56

Figure 1-57

L Click OK on all open dialog boxes.

On the map, the park labels are visible against the basemap (Figure 1-58).

Figure 1-58

Figure 1-59

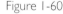

The new symbology is a big improvement, but the labels would look better still if the text were stacked instead of strung out in a line. To make this change, we'll switch from ArcMap's default labeling system to its advanced system.

Ⓜ From the main menu, choose View → Data Frame Properties.

Ⓝ In the Data Frame Properties dialog box, if necessary, click the General tab.

The "data frame" refers generally to the map view. Data frame properties apply to the map view as a whole or to all the layers within it equally. You'll work more with data frame properties in Lessons 3 and 8.

Ⓞ At the bottom of the dialog box, click the Label Engine drop-down arrow and choose Maplex Label Engine (Figure 1-59).

ArcMap has two labeling "engines." Both place labels on the map automatically according to a set of preferences that you can adjust. (For example, you might want labels placed to the left of features rather than to the right.) The Maplex Label Engine has a more robust set of preferences than the Standard Label Engine. We won't explore label functionality very deeply in this book, but the Maplex Label Engine offers one immediate advantage: it stacks label text automatically.

Ⓟ Click OK on the Data Frame Properties dialog box.

On the map, the park labels are stacked, keeping them closer to their features (Figure 1-60).

Figure 1-60

6) **Set a scale range for the labels.** The park labels look good at this fairly large scale, where there's room to accommodate them. Let's see what happens when we zoom out.

Ⓐ Zoom to the City of Los Angeles bookmark.

At this smaller (zoomed-out) scale, the labels overwhelm the map (Figure 1-61). We could just turn them off, of course, whenever they seem to be too crowded. A nicer solution is to make their visibility dependent on the map scale.

Figure 1-61

Ⓑ Open the layer properties for the *Parkland* layer. Click the Labels tab if necessary.

Ⓒ Near the bottom of the Labels tab, click Scale Range.

Ⓓ In the Scale Range dialog box, click the "Don't show labels when zoomed" option.

Ⓔ Click in the "Out beyond" box and type **40,000**.

Whenever the map scale crosses the 1:40,000 threshold, the park labels will turn off automatically.

Ⓕ Compare your Scale Range dialog box to Figure 1-62 and click OK. Click OK again on the Layer Properties dialog box.

Figure 1-62

The park labels (that is, the ones that you set) disappear from the map.

Ⓖ Zoom to the Dodger Stadium bookmark.

As long as the map scale is larger than 1:40,000 (which it should be), the labels show up again.

You've probably noticed that some parks, such as Elysian Park and Echo Park, are labeled in both the *Reference* and *Parkland* layers. We can set the scale range of the *Reference* layer reciprocally to the *Parkland* labels.

Ⓗ Open the layer properties for the *Reference* layer. Click the General tab, if necessary.

Figure 1-63

This layer, being a remotely hosted map service, has fewer properties you can set than a layer whose source data is on your computer.

Ⓘ In the Scale Range area, click the "Don't show layer when zoomed" option.

Ⓙ Click in the "In beyond" box and type **40,001**.

Ⓚ Compare your dialog box to Figure 1-63 and click OK.

The labels from the *Reference* layer should disappear from your map. (Most of these labels are actually for places other than parks, but we'll accept their loss at large scales.)

Ⓛ Zoom back and forth across the 1:40,000 scale threshold to test your settings.

At scales of 1:40,000 or larger, the *Reference* layer appears with a gray check box in the Table of Contents (Figure 1-64). This tells you that the layer is turned on, but is set not to show at the current map scale.

Figure 1-64

Scale ranges

As you've just seen, scale ranges can be set both for a layer's labels and for the layer itself. In this case, the distinction was blurred because the *Reference* layer is composed of labels. But layers that are composed of features, such as parks or rivers, can have their scale ranges set in exactly the same way. When you design maps to be viewed at multiple scales, you normally want layers representing detailed data, such as buildings or utility lines, to be visible only at large (close-up) scales. You may want your map to include multiple copies of a layer, each with a different scale range and unique symbology. For example, one layer might represent trees with a generic point symbol at medium scales. A second layer might represent the same trees with a more detailed and realistic symbol at large scales.

7) **Dim the basemap.** Without doing anything further to the labels, we can emphasize them a little more by dimming the imagery basemap. To do that, we'll add a toolbar to the interface.

Ⓐ From the main menu, choose Customize → Toolbars → Effects.

The Effects toolbar is added (Figure 1-65). ArcMap has a lot of special-purpose toolbars, and we'll use several of them in this book.

Figure 1-65

Ⓑ Drag the toolbar to a position you like.
  ▷ You can dock it on the top, bottom, or either side of the ArcMap window, or leave it undocked anywhere on your screen.

Ⓒ On the Effects toolbar, click the drop-down arrow and choose the *Basemap* layer.

Ⓓ Click the Adjust Dim Level button ◇ to open a slider bar.

Ⓔ Set the Dim Level to 10% (Figure 1-66).

On the map, the imagery fades slightly.

Figure 1-66

8) **Add a layer of census tracts.** The U.S. Census Bureau gathers socioeconomic data about households and aggregates it by various geographic units. One of these units is a census tract, which is a relatively small subdivision of a county.

Ⓐ Open the Catalog window.

Ⓑ Under the SourceData folder, expand the census folder.

Ⓒ Drag and drop *tracts.shp* onto the map.

The *tracts* layer is added to the top of the Table of Contents. The tracts cover everything except the *Parkland* labels.

Ⓓ In the Table of Contents, right-click the *tracts* layer and choose Zoom To Layer.

The census tract data covers Los Angeles County, including the islands of Catalina and San Clemente.

Ⓔ Zoom to the City of Los Angeles bookmark.

Ⓕ Drag the *Reference* layer to the top of the Table of Contents.

The *tracts* layer is made up of contiguous polygons that are reminiscent of a jigsaw puzzle (Figure 1-67). Since each piece of the puzzle represents a different set of living, breathing human beings, it's natural that each tract's attribute values for population, income, age, and so on, are different.

Figure 1-67

You may see a different set of labels, depending on your map scale. Map scale is affected by things like the size of your ArcMap window and the resolution of your monitor.

9)   **Open the attribute table.** Let's see what attributes the table has.

Ⓐ In the Table of Contents, right-click the *tracts* layer and choose Open Attribute Table.

Ⓑ Scroll across the table and look at the field names.

The table has some identification codes, followed by population attributes and breakdowns of ethnicity, sex, age, household status, and housing. Some of the field names are fairly easy to interpret, others less so. In Lesson 2, we'll see how to get more information about our information by accessing metadata.

Ⓒ Locate the POP10_SQMI field (Figure 1-68).

Figure 1-68

FIPS	POP2000	POP00_SQMI	POP2010	POP10_SQMI
06037101110	4500	10227.3	4731	10752.3
06037101122	3176	3113.7	3664	3592.2
06037101210	6066	24264	5990	23960
06037101220	3028	11214.8	3363	12455.6
06037101300	3974	4096.9	4199	4328.9
06037101400	3760	1579.8	3903	1639.9
06037102103	1705	3552.1	1740	3625
06037102104	2974	4248.6	3648	5211.4

|◄ ◄   1 ► ►| ☷ ▤ | (0 out of 2343 Selected)

tracts

This attribute stores population density (people per square mile) for the year 2010. Since one of our criteria is to locate the park in a densely populated area, this is relevant information. The attribute doesn't tell us what value should be a threshold for "high density," but it does give us a way to start making patterns on a map.

❶ Right-click the POP10_SQMI field name and choose Statistics (Figure 1-69).

Figure 1-69

The Statistics of tracts window displays summary statistics for the field. The lowest value is 0 (at least one tract must be unpopulated) and the highest is 98,150. The frequency distribution chart, or histogram, on the right shows you that most of the values are between 0 and about 25,000. The remaining values spread out in a long tail.

❷ Close the Statistics of tracts window.

❸ Close the table.

10) Symbolize census tracts by population density. Symbolizing a layer by an attribute, also called thematic mapping, allows us to see how values are spatially distributed.

❹ Open the layer properties for the *tracts* layer and click the Symbology tab.

By default, all features in a layer have a single symbol (Figure 1-70). That's why all your census tracts are purple, or whatever color they happen to be.

❺ In the Show box, click Quantities.

Figure 1-70

Under Quantities, four methods for symbolizing numeric attributes are listed. The default Graduated colors method will give each feature in the *tracts* layer a color that reflects its numeric value for a chosen attribute.

C In the Fields area, click the Value drop-down list and choose POP10_SQMI (Figure 1-71).

Figure 1-71

A lot is going on here. The values for the POP10_SQMI attribute, which range from 0 to 98,150, are divided into five classes. The starting and ending values for each class are calculated by a "natural breaks" algorithm that separates clumps in the data. That's why the range of values is different from class to class and why classes break at seemingly arbitrary numbers. Each class is associated with a symbol in a color ramp (probably yellow to dark red, but yours may be different if you happened to change it).

D Click Apply. Move the Layer Properties dialog box out of the way.

On the map, the tracts are symbolized by population density (Figure 1-72).

11) Change the classification. Quantitative symbology is flexible, and you can present data in many ways. Because all we want right now is a general sense of viable areas for our project, and because we're going to look at a couple of variables together, we should keep our presentation simple.

A In the Layer Properties dialog box, click the Classes drop-down arrow and choose 3.

Figure 1-72

❸ Click Classify to open the Classification dialog box
(Figure 1-73).

Figure 1-73

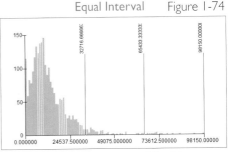

Equal Interval   Figure 1-74

Classes are evenly spaced. In this case,
almost all the records fall in the first class.

The histogram shows you the distribution of values in relation to the
current class breaks. We can change the algorithm used to set class
breaks, or make manual adjustments as desired.

❸ Click the Method drop-down arrow and choose Equal Interval
(Figure 1-74).

❹ Now set the Classification Method to Quantile.

Quantile

Each class has an equal number of
records. Class spacing is very different.

With three classes, we'll easily be able to see high, medium, and low
values. The Quantile method guarantees that an equal number of
tracts will fall into each class. It should be noted that there are no
inherently good or bad ways to classify data—different classifications
may be more or less appropriate to the purpose of your map and the
background knowledge of your audience.

❺ In the Break Values box to the right of the histogram, click the
first class break point (7715).

The value becomes editable. At the bottom of the dialog box, a
message tells you that there are 772 elements (in this case, census
tracts) in this class.

❻ Replace the highlighted value with **8,000** and press Enter.

The second class break point is selected and editable.

❼ Replace the highlighted value (14196) with **16,000**.

❽ Click in some white space in the Histogram window to stop
editing class breaks (Figure 1-75).

Figure 1-75

At the top of the dialog box, the classification method has been reset to Manual because we've changed the class breaks. The histogram is updated, too. We no longer have a pure quantile classification, but having our classes break at round numbers makes intuitive sense.

❶ Click OK on the Classification dialog box.

12) Change the symbology. We'll make some changes to the symbology as well.

Ⓐ On the Layer Properties dialog box, click the Color Ramp drop-down arrow (or click on the ramp itself).

Ⓑ Scroll up and down to see the choices, then click the ramp again to close the list.

Ⓒ Right-click the color ramp. On the context menu, click Graphic View to uncheck it (Figure 1-76).

Figure 1-76

Ⓓ Click the ramp's name. In the drop-down list, click the Red Light to Dark ramp to select it (Figure 1-77).

Ⓔ Right-click the ramp name and choose Graphic View to show the ramps by color again.

Figure 1-77

Underneath the color ramp is a box with three columns:

- The Symbol column shows the symbol for each class.
- The Range column shows the range of values for each class.
- The Label column shows how each symbol will be described in the Table of Contents. (By default, the label matches the range.)

Ⓕ Click the Symbol column heading and choose Properties for All Symbols.

Ⓖ In the Symbol Selector, click the Outline Color button. On the color palette, choose No Color.

Ⓗ Click OK on the Symbol Selector.

We're taking away the outlines because we don't need to see the tract boundaries on the map. For now, we're interested in them as areas, not specifically as tracts.

Ⓘ In the Label column, click on the first label (0.000000 - 8000.000000) to make it editable. Type **Low** and press Enter.

Ⓙ Replace the second label with **Medium**. Press Enter.

Ⓚ Change the third label to **High**. Click outside the edit box to commit the edit.

Ⓛ Compare your settings to Figure 1-78 and click Apply.

Figure 1-78

Ⓜ In the Layer Properties dialog box, click the Display tab. In the Transparent box at the top, replace the value 0 with **50** (Figure 1-79).

Ⓝ Click OK on the Layer Properties dialog box.

Ⓞ In the Table of Contents, drag and drop the *tracts* layer underneath the *Los Angeles* layer.

Figure 1-79

On the map, we can now see where population is concentrated along the river, and we can see it in relation to existing parks (Figure 1-80).

Figure 1-80

13) Measure distance from the river to parks. Making a few measurements will improve our ability to estimate distance on the map and will give us a better intuitive sense of how close to the river the new park should be.

- **A** Zoom to the Dodger Stadium bookmark.
- **B** Pan the map so that a number of parks are in the view. Feel free to zoom in or out.
- **C** On the Tools toolbar, click the Measure tool 📏 to open the Measure dialog box.
- **D** In the row of tools at the top of the Measure dialog box, make sure the Measure Line tool is selected (Figure 1-81).
- **E** Click the Choose Units button and choose Distance → Miles.

Confirm that
Measure Line
is selected.

Click here to
choose the
measurement
unit.

Figure 1-81

On the map, the mouse pointer changes to a ruler with inscribed crosshairs.

- **F** Move the mouse pointer over a park, such as Cypress Park (northeast of Dodger Stadium on the east side of the river).

Figure 1-82

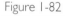

The mouse pointer "snaps" to the park boundary. You can probably feel the effect in the mouse movement. It's confirmed on-screen by a light gray square near the mouse pointer and a message such as "Parkland: Vertex" or "Parkland: Edge" (Figure 1-82).

- **G** Click to start a measurement.
- **H** Move the mouse pointer (you don't have to drag) to the river.

The mouse pointer snaps to the river feature. The measurement result is displayed in the Measure dialog box (Figure 1-83).

- **I** Double-click to end the measurement.
- **J** Click on another park and measure its distance to the river.

Figure 1-83

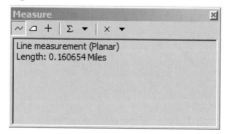

The new result replaces the previous one in the Measure dialog box.

- **K** Measure the distances from a few more parks.

Your result will depend on where you start and end your measurement.

Cypress Park, Elysian Valley Rec Center Park, and Downey Playground are close to the river. Elyria Canyon Park, a little over three quarters of a mile away (at its nearest edge), stretches the notion of proximity. Bear in mind that these measurements are straight-line distances, not distances along streets.

Ⓛ Close the Measure dialog box.

Although the dialog box is closed, the tool stays active until another tool is selected.

Ⓜ Zoom to the City of Los Angeles bookmark.

14) Add a layer of census block groups. Another requirement for the new park is that it be located in a lower-income neighborhood. To symbolize an income attribute, we have to add another layer to the map. (It's not that income data isn't reported at the census tract level, it's just that our particular *tracts* layer doesn't happen to include it.)

Ⓐ Open the Catalog window.

The same census folder that contains *tracts.shp* also has a dataset called *block_groups.shp*.

Ⓑ Drag and drop *block_groups.shp* on the map.

In the Table of Contents, the new layer is added above the other layers. Like the *tracts* layer, the *block_groups* layer covers Los Angeles County. And like the census tracts, the block group polygons resemble a jigsaw puzzle. Block groups are another Census Bureau statistical unit: they're smaller than tracts and nest inside them. We'll talk more about census geography in Lesson 2.

Ⓒ Open the attribute table for the *block_groups* layer and scroll across the attributes.

Most of these attribute names are cryptic. One of the last fields in the table is called MEDHINC_CY (Figure 1-84). For now, take it on faith that this stands for median household income.

RENTER_CY	VACANT_CY	POP18UP_CY	MEDAGE_CY	MEDHINC_CY	AVGHINC_CY	AGGHINC_CY	PCI_CY
266	17	761	42.1	29782	49196	19924561	21788
70	5	1400	32	60122	78449	27849561	16534
32	5	647	34.4	56282	69003	14421723	17566
672	62	1501	29.8	38276	55837	38416014	19314
64	23	1708	43.1	94370	119792	75109361	36720
126	16	710	34.5	74676	117693	39191826	39086
746	56	2659	37.6	42003	58348	61499122	19699

(0 out of 6421 Selected)

Figure 1-84

Ⓓ Right-click the MEDHINC_CY field name and choose Statistics.

The Statistics of block_groups window tells you that the lowest value in the field is 0 and the highest is 200,001. Median value is a midpoint. For each block group, half the households earn more than the median income and half earn less.

E Close the Statistics of block_groups window.

F Close the attribute table.

G Drag the *Reference* layer to the top of the Table of Contents.

15) Symbolize census block groups by median household income. If we symbolize the *block_groups* layer with graduated colors, we won't be able to evaluate income and population density at the same time. Instead, we'll represent each block group's median household income as a point drawn inside the block group polygon. The point sizes will be graduated according to the income value.

A Open the layer properties for the *block_groups* layer and click the Symbology tab.

B In the Show box, click Quantities. Under Quantities, click Graduated Symbols (Figure 1-85).

C In the Fields area, click the Value drop-down list and choose MEDHINC_CY.

D Click the Classes drop-down arrow and choose 3.

E Click Classify to open the Classification dialog box.

By coincidence, the first break point is already a nice round number (50,000), so we don't have to alter it.

F In the Break Values box to the right of the histogram, click the second class break point (94370) to make it editable.

G Replace the highlighted value with **100,000** and press Enter.

H Click in some white space in the Histogram window to stop editing class breaks (Figure 1-86). Click OK on the Classification dialog box.

Figure 1-85

Figure 1-86

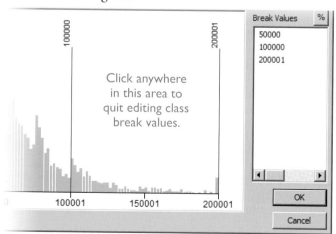

ⓘ In the Layer Properties dialog box, click Template (underneath the Classify button) to open the Symbol Selector.

ⓙ In the scrolling box of symbols, click Circle 2 to select it.

ⓚ Click the Color button and change its color to Tourmaline Green (Figure 1-87).

ⓛ Click OK on the Symbol Selector.

This sets the color and shape of the symbols that will be used to represent income values.

ⓜ Click Background (underneath Template) to open the Symbol Selector again.

ⓝ Change the fill color to No Color, then change the outline color to No Color. Click OK.

This makes the block group polygons themselves invisible. All we'll see are the income dots spread around the map.

ⓞ In the Symbol Size boxes, replace the "from" value with **8** and the "to" value with **24**.

ⓟ In the Label column, click on the first label (0 - 50000) to make it editable. Type **Low** and press Enter.

ⓠ Replace the second label with **Medium**. Press Enter.

ⓡ Change the third label to **High**. Click outside the edit box to commit the edit.

ⓢ Compare your settings to Figure 1-88. Click Apply and move the Layer Properties dialog box out of the way.

At the present scale, the symbols overwhelm the map (Figure 1-89).

Figure 1-87

Figure 1-88

Figure 1-89

Your view may look different depending on the size of your computer screen and ArcMap window, but the effect should be similar.

16) Set a scale range for the *block_groups* layer. Earlier, we set a maximum scale value for the *Reference* layer. Here, we'll set a minimum value for the *block_groups* layer.

Ⓐ On the Layer Properties dialog box, click the General tab.

Ⓑ In the Scale Range area, click the "Don't show layer when zoomed" option.

Ⓒ Click in the "Out beyond" box and type **100,000** (Figure 1-90) and click OK.

Figure 1-90

Figure 1-91

The symbols disappear from the map. In the Table of Contents, the layer's check box indicates that the layer is turned on but is not visible at the current map scale (Figure 1-91).

Ⓓ Zoom to the Dodger Stadium bookmark.

Ⓔ Click the Fixed Zoom Out button ⬚ several times until some of the Medium and High symbols begin to appear.

Now we can start to get a general sense of household income and population density along the river, and look at these variables in relation to park locations (Figure 1-92).

Figure 1-92

17) Search for likely park areas. Clearly, we're taking an incomplete initial look at a complex problem. We haven't considered all the requirements (for example, the presence of children). We're not making any exact measurements of distance. Our data classifications are casual: we don't yet have a good reason to say what values should count as high population density or low median household income in the context of our project. Nevertheless, we can form some meaningful impressions. We won't be able to say of an area that a park should *definitely* go there, but we might be able to identify likely and unlikely areas. Later, it will be interesting to see how well these impressions are borne out by analysis.

Figure 1-93

Ⓐ On the Standard toolbar, in the Map Scale box, highlight the current value. Type **40,000** and press Enter.

Ⓑ Pan south to where the river crosses the city boundary.

Ⓒ In the Table of Contents, drag the *Los Angeles River* layer to the position just below the *Reference* layer (Figure 1-93).

We'll follow the river to its source, marking good areas along the way. A really good area would have these properties:

- High population density (dark red)
- Low median household income (small green dot)
- No existing park nearby
- Close to the river

Ⓓ Pan slowly north.

Less than a mile north of the city limits, on the east side of the river, are a couple of tracts—one dark red and one medium red—with small green dots (Figure 1-94). They're pretty close to the river, and even though there are some parks in the general vicinity, it's probably worth marking the area.

Figure 1-94

E From the main menu, choose Customize → Toolbars → Draw.
The Draw toolbar is added.

F Dock the toolbar or float it, whatever you like best.

G On the Draw toolbar, click the Drawing drop-down menu and choose Default Symbol Properties at the bottom.

H On the Default Symbol Properties dialog box, click the Fill button to open the Symbol Selector.

I Set properties in the Symbol Selector as follows:
- Set the fill color to No Color.
- Set the outline width to 3.
- Set the outline color to Black.

J Click OK on the Symbol Selector.

K Compare your Default Symbol Properties dialog box to Figure 1-95 and click OK.

Figure 1-95

L On the Draw toolbar, click the drop-down arrow next to the Rectangle tool. From the list of tools, choose the Ellipse tool.

M On the map, click and drag a box over the area of interest. Release the mouse button to draw the ellipse (Figure 1-96).

You can reshape the ellipse by dragging its selection handles. You can move it by clicking and dragging it. The exact size and position of the shape doesn't matter as long as you mark the general area.

N Click off the graphic to unselect it.

Figure 1-96

Graphics

Map graphics, such as these created with tools on the Draw toolbar, are different from features. Conceptually, features represent geographic objects while graphics tend to be visual aids. Technically, graphics are usually stored in map documents as annotation, not in feature classes on disk. (You'll learn more about this in Lesson 8.) They aren't maintained as layers in ArcMap, nor do they have attribute tables, but they can be managed from the Data Frame Properties dialog box (Annotation Groups tab). It is possible to convert features to graphics and graphics to features.

18) Continue searching for park areas. We'll keep following the river and looking for likely places for a park.

Ⓐ Pan north.

A little farther north, again on the east side of the river, is a medium-density, lower-income area with just one park (Pecan Playground) in the general vicinity. We'd prefer a high-density neighborhood, but since we don't yet have specific criteria for our analysis, we want to be inclusive rather than exclusive in our assessments. It's up to you whether to mark this area or not—all that really matters is that you start to gain a sense of the study area.

Ⓑ On the Drawing menu, click the Ellipse tool and draw an ellipse around the area (Figure 1-97).

Ⓒ Unselect the graphic and keep panning along the river.

The area opposite Dodger Stadium, on the east side of the river, is medium-density, lower-income, and park poor.

Ⓓ Draw an ellipse around this area (Figure 1-98).

You may find that the shape of the ellipse doesn't fit the area just the way you want.

Ⓔ On the Draw toolbar, click the Rotate button ↻.

Ⓕ Place the mouse pointer over the graphic, then click and drag to rotate the graphic. Orient the ellipse to fit the area better.

Ⓖ On the Draw toolbar, click the Select Elements tool ➤.

Ⓗ Resize and reshape the ellipse by dragging selection handles.

You may need to switch back and forth between the Rotate and Select Elements tool to make the ellipse look the way you want (Figure 1-99). It's not art, so don't spend too much time on it.

Figure 1-97

Figure 1-98

Figure 1-99

ⓘ Unselect the graphic and continue panning along the river. Mark areas that seem promising to you and reshape the ellipses as needed.

As you navigate around Griffith Park, remember that the yellow line marks the city limits. The areas north and east of the park aren't part of Los Angeles and shouldn't be considered. Once you get around this park, you may find fewer likely areas. But remember, there's no right or wrong answer for this exercise.

ⓙ After adding the last ellipse, from the main menu, choose Edit → Select All Elements.

ⓚ Again from the main menu, choose Edit → Zoom To Selected Elements.

ⓛ Click anywhere on the map to unselect the graphic elements (Figure 1-100).

ⓜ Save the map document.

Figure 1-100

19) Export the map. We'll export this view of the map to an image in TIFF format. When we finish the analysis in Lesson 6, we'll add the image and compare our guesses to the actual results.

Ⓐ From the main menu, choose File → Export Map.

Ⓑ At the top of the Export Map dialog box, click the Home 🏠 button to navigate to the MapsAndMore folder.

Ⓒ In the Save as type box, click the drop-down arrow and choose TIFF (*.tif).

Ⓓ Replace the file name with **Lesson1Predictions**.

Ⓔ Under Options, click the Format tab. At the bottom, check the box to write GeoTIFF tags.

GeoTIFF tags embed coordinate system information into the image file. The output will be an image that can be used in documents such as presentations or reports, but that will also align spatially with layers in ArcMap.

Ⓕ Compare your Export Map dialog box to Figure 1-101 and click Save.

The Export Map dialog box closes by itself.

20) Exit ArcMap. Toolbars that you leave on the interface will remain there in subsequent sessions. There's nothing wrong with that, but too many toolbars can become a distraction.

Ⓐ From the main menu, choose Customize → Toolbars → Draw.

The Draw toolbar is removed.

Ⓑ Remove the Effects toolbar in the same way.

Ⓒ From the main menu, choose File → Exit.

At this point, you already have a basic working knowledge of ArcMap. You know the problem to be solved and you have a sense of the study area. You've worked with some of the available data for the project, but you haven't taken a careful, systematic look at it. That will be the focus of the next lesson.

Figure 1-101

Symbology methods

The default **Single Symbol** method (under Features) assigns the same symbol to all features in a layer. No attribute is symbolized.

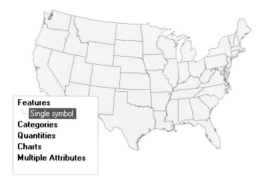

Under the **Categories** heading are methods for symbolizing descriptive attributes. Here, the Unique Values method assigns different line widths to a roads layer according to whether the roads are freeways, highways, or streets.

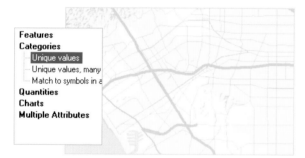

Under the **Quantities** heading are methods for symbolizing numeric data. In this example, census tracts are symbolized by population density. Low-density tracts are yellow, medium-density are orange, and high-density are brown.

The **Charts** heading has methods to compare numeric attributes by chart symbols. These pie charts compare percentages of owner-occupied homes (red), renter-occupied homes (blue), and unoccupied homes (yellow) for Southern California cities.

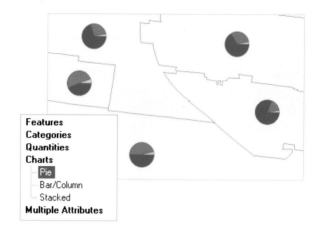

The **Multiple Attributes** heading lets you symbolize descriptive and numeric attributes together. Here, the descriptive attribute is vacancy. Vacant lots are orange; developed lots are pink. The numeric attribute is land value. A small dot in a vacant lot means the value is low; a large dot means the value is high.

Lesson
2 Preview data

WE NEED THE RIGHT DATA TO SOLVE

the problem. We explored the study area in Lesson 1; now we can proceed more systematically. What data do we have? How useful is it? Is there data that we need but don't have? Has the problem been stated clearly enough for us to know what data we need?

Acquiring, evaluating, and organizing data is a big part of an analysis project. This book doesn't fully re-create the complexity of the real world because we've provided all the basic data required. On the other hand, much of the data isn't project-ready, and that does reflect the real world of GIS.

The first thing we'll do in this lesson is draw up a planning document to help us keep our tasks in focus. We'll use this document to list the guidelines for the new park and translate them into specific needs for spatial and attribute data.

After we itemize our data requirements in general terms (park data, river data, and so on), we'll take stock of our source data and investigate its spatial and attribute properties. We'll also familiarize ourselves with metadata, which is the data we have about our data. Before you decide to use a particular dataset, you may want to know things like who made the data, when, and to what standard of accuracy.

Once we have a better working knowledge of our data, we'll reframe the problem statement. GIS is a quantitative technology: we can't analyze a problem until it's been stated in measurable terms. Wherever we find the city council's guidelines to be vague, we'll replace them with hard numbers.

Lesson Two roadmap

You are here:

What you'll do in this lesson:

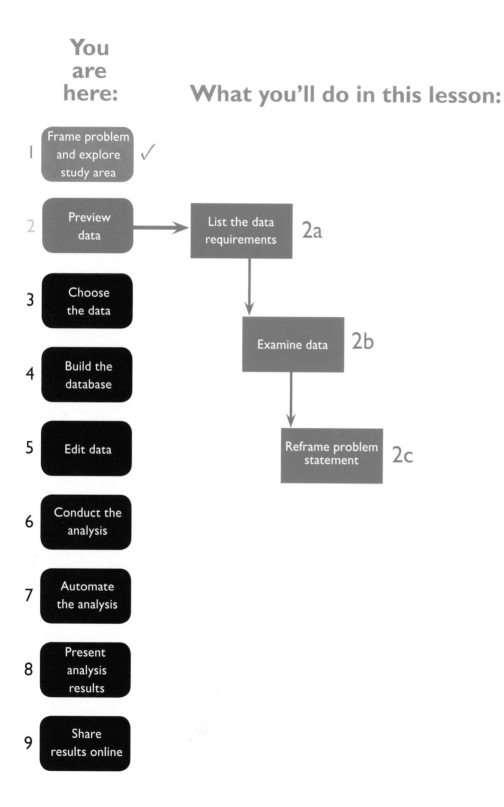

1 Frame problem and explore study area ✓

2 Preview data → List the data requirements 2a

↓

Examine data 2b

↓

Reframe problem statement 2c

3 Choose the data

4 Build the database

5 Edit data

6 Conduct the analysis

7 Automate the analysis

8 Present analysis results

9 Share results online

Exercise 2a: List the data requirements

We need to relate the guidelines for the new park to data requirements for the project.

1) Open the data requirements table. We've made a table in advance to keep track of our requirements. It's an informal document, but it will still be helpful.

- **A** Open Windows Explorer and navigate to C:\UGIS\ParkSite\MapsAndMore.
- **B** Double-click the file *DataRequirementsTable.doc* to open it in Microsoft Word (Figure 2-1).
  - ▷ If you don't have Microsoft Word, open the .rtf version of the document in another application, or print the .pdf version and fill it out with a pencil.

Figure 2-1

#	REQUIREMENT	DEFINED AS	SPATIAL DATA	ATTRIBUTE DATA	DATA SET	PREPARATION
1	land parcel		parcels			
2	vacant					
3						
4						
5						
6						
7						
8						
9						
10						
11						
12						
13						
14						
15						

Pieces of land, legally surveyed, described, and recorded with a county or other administrative body are called *parcels*.

2) List the requirements. In this step, we'll review the city council's guidelines (see next page) and describe in a general way the data needed to satisfy them. We'll get to the specifics of choosing datasets in Lesson 3.

The first guideline was to find a vacant piece of land at least one acre in size. We can break this down into three requirements:

- land parcel
- vacancy
- size

The requirement for a land parcel is already listed in the table. We need spatial data representing parcels so that we can see candidate sites on the map.

The second requirement is vacancy, which is a characteristic, or attribute, of a parcel. In a GIS dataset, vacancy is often listed with other descriptions of land use (commercial, residential, industrial, and so on). In general terms, then, we're looking for a land use attribute.

Figure 2-2

Ⓐ In Row 2 of the table, under ATTRIBUTE DATA, type (or write) **land use**.

The third requirement is that the park be one acre or larger. Like vacancy, acreage is an attribute, although it is one that can be calculated by the software. Since ArcMap can convert one unit of area to another, we don't even have to start with acres—any measurement of parcel size will suffice.

Ⓑ In Row 3, under REQUIREMENT, type **one or more acres**. Under ATTRIBUTE DATA, enter **area**.

The second guideline is that the park be within the LA city limits (Figure 2-2). This sounds like spatial data and we'll treat it that way for now. (It could be an attribute, too, because a field in a table might store the name of the city in which each parcel is recorded.)

Ⓒ Fill in Row 4 as you think it should look, then check Figure 2-3.

#	REQUIREMENT	DEFINED AS	SPATIAL DATA	ATTRIBUTE DATA	DATA S
1	land parcel		parcels		
2	vacant			land use	
3	one or more acres			area	
4	within city limits		cities		
5					
6					

Figure 2-3

The third guideline is that the park be as close as possible to the Los Angeles River.

Ⓓ In Row 5, for the requirement, put **near LA river**. Under SPATIAL DATA, put **rivers**.

With spatial datasets of parcels and rivers, we can use ArcMap tools to measure the distance from any given parcel to the river.

The fourth guideline is to put the park away from existing parks.

Ⓔ Fill out Row 6 as you think it should look.

The fifth guideline also needs to be broken down. We need

- A neighborhood (spatial data) with
- High population density (attribute data) and
- Lots of children (attribute data)

Neighborhoods tend not to have formal boundaries, so we're probably not going to find them as such in a spatial dataset. As a proxy, or substitute, we'll use a set of small, standardized areas defined by the U.S. Census Bureau: either the tracts or block groups we looked at in Lesson 1.

F In Row 7, enter **in a neighborhood** as the requirement. Enter **census unit** for the spatial data.

G In Row 8, enter **densely populated** for the requirement and **population density** for the attribute data.

H In Row 9, enter **lots of kids** for the requirement. For the attribute data, enter **age**.

The sixth guideline is that the park go in a lower-income neighborhood. We don't need to repeat the spatial requirement for a neighborhood.

I In Row 10, enter **lower income** for the requirement and **income** for the attribute data.

The last guideline is to serve as many people as possible. For this we need a population attribute.

J In Row 11, enter **serving the most people** as the requirement and **population** as the attribute data (Figure 2-4).

#	REQUIREMENT	DEFINED AS	SPATIAL DATA	ATTRIBUTE DATA	DATA SET
1	land parcel		parcels		
2	vacant			land use	
3	one or more acres			area	
4	within city limits		cities		
5	near LA river		rivers		
6	away from parks		parks		
7	in a neighborhood		census unit		
8	densely populated			population density	
9	lots of kids			age	
10	low income			income	
11	serving the most people			population	
12					

Figure 2-4

Eventually, we want to make a map of potential sites, and we may need some data for cartographic purposes. For example, political boundaries and roads put maps in a familiar context. Physical relief creates texture and imagery provides realistic detail.

K In Rows 12 to 15, enter **final map** for the requirement. Under SPATIAL DATA, list the examples just mentioned (Figure 2-5).

L Save and minimize the table. We'll continue to use it.

#	REQUIREMENT	DEFINED AS	SPATIAL DATA	ATTRIBUTE DATA	DATA SET	PREPA
1	land parcel		parcels			
2	vacant			land use		
3	one or more acres			area		
4	within city limits		cities			
5	near LA river		rivers			
6	away from parks		parks			
7	in a neighborhood		census unit			
8	densely populated			population density		
9	lots of kids			age		
10	low income			income		
11	serving the most people			population		
12	final map		political boundaries			
13	final map		roads			
14	final map		relief			
15	final map		imagery			

Figure 2-5

Exercise 2b: Examine data

Let's see what data we actually have on hand. To do this, we'll work in ArcCatalog, the data management application of ArcGIS for Desktop. In Lesson 1, you used the Catalog window, which is basically ArcCatalog tailored to work inside ArcMap. The Catalog window is great for going back and forth between your map and your data (which is what you do most of the time). In this lesson, however, we won't be doing any mapping, and the stand-alone ArcCatalog application will be more convenient to work with.

1) **Start ArcCatalog.** We'll open the ArcCatalog application.

Ⓐ Click the Windows Start button and choose All Programs → ArcGIS → ArcCatalog 10.1.

▷ If you have a shortcut on your desktop, you can double-click the ArcCatalog 10.1 icon 🔷 instead.

ArcCatalog opens (Figure 2-6). The Catalog tree, on the left, is the counterpart of the Catalog window in ArcMap. On the right are three tabs.

- The Contents tab shows file information.
- The Preview tab lets you explore spatial and attribute data.
- The Description tab displays metadata (data about data).

Figure 2-6

2) **Expand the folder connection.** At the top of the Catalog Tree is a folder called Folder Connections.

Ⓐ In the Catalog Tree, expand Folder Connections by clicking its plus sign.

Underneath it, you should see the C:\UGIS folder connection that you made from ArcMap in Lesson 1.

▷ If you don't have this folder connection, make it now. On the Standard toolbar, click the Connect To Folder button ![]. Browse to C:\UGIS and click OK.

Ⓑ Expand the C:\UGIS folder.

Ⓒ Expand the ParkSite folder and the SourceData folder below it.

Ⓓ Under SourceData, expand everything you can.

It's a long list of items (Figure 2-7). With the Catalog Tree docked in ArcCatalog, you may have to scroll down or maximize the application to see everything. Each item is a piece of geographic data or a data container. The icons signify the type of data, as illustrated in the topic *Representing the real world as data*.

3) **Survey the data.** Under the SourceData folder are three folders 📁 and a geodatabase 🛢.

- The census folder contains three feature classes of census data in shapefile format.
- The City of LA folder contains three shapefiles and a stand-alone table in dBASE (.dbf) format.
- The ParkData folder contains a shapefile and a raster dataset in TIFF format.
- The ESRI geodatabase contains ten feature classes in geodatabase format. The feature classes are thematically organized in containers called feature datasets 🗂.

In the next several steps, we'll preview a lot of this data to make sure we have the features and attributes we listed in the data requirements table.

Figure 2-7

Acquiring data

In any GIS project, acquiring good data is a big part of your job. Datasets representing many types of geography are available from ArcGIS Online. You can access this data from a web browser (www.arcgis.com) or directly from ArcMap (File → Add Data → Add Data From ArcGIS Online). You can search for particular types of data by keyword; you can also search for ArcGIS Online groups, such as Esri Maps and Data, that curate a variety of authoritative content. Spatial data is also widely available from government agencies, educational institutions, and commercial vendors. All of these sources may supplement data collected and managed by your own organization.

Representing the real world as data

How would you create an information system to organize and manage the huge variety of geographic stuff in the world? One approach is to think of all that stuff in terms of discrete objects.

The discrete object view of the world

If you conceive geography in terms of objects, you can sort these objects by similarities. Shape is a fundamental sorting principle: every object can be drawn—in two dimensions—as either a point, a line, or a polygon. Theme, or type, is another principle: every object can be classified as a school, a road, a park, or something else.

Applying these sorting principles of shape and theme, we can come up with collections of things we would recognize on a map: schools represented as points, roads represented as lines, parks represented as polygons, and so on.

Points, lines, and polygons representing geographic objects on a map.

Each object in a collection has a unique location, specified by a pair of spatial coordinates (for points) or a list of coordinate pairs (for lines and polygons).

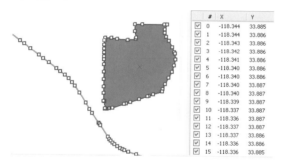

A polygon (with an x indicating its centroid) and the list of coordinate pairs that defines its location.

Besides a unique location, every object has a set of facts that pertain to it: a name, a description, or whatever bits of information have been gathered about it. These facts are the object's attributes.

FID	Shape ^	PARK NAME	CATEGORY	ADDRESS	TELEPHONE	HOURS
0	Polygon	Los Angeles	State Historic Park	1245 N. Spring Street Los	213-620-6152	8 AM - sunset
1	Polygon	Rio de Los Angeles	State Recreation Area	1900 San Fernando Road	213-620-6152	9 AM - 10:30 PM
2	Polygon	Vista Hermosa Park	Local Park or Recreation Area	100 N. Toluca Street Los	213-250-3578	8 AM - sunset

In ArcGIS, a collection of such objects—with a common shape, common theme, and common attributes—is called a feature class. An individual object in the collection is a feature. The feature class is the basic storage unit for GIS data created according to the discrete object view of the world, commonly called the vector data model.

Feature classes can be stored in various file formats, notably the geodatabase and the shapefile. The geodatabase format is newer and more highly developed.

Feature classes in geodatabase format (left) and shapefile format (right). Different icons indicate polygon, line, and point data.

Representing the real world as data (continued)

The continuous surface view of the world

Although it's a powerful model, the discrete object view is not an intuitive way to think of certain kinds of geographic information, like elevation or temperature, that don't have shapes or boundaries and that cover the world everywhere. It's quite possible to represent these phenomena as features (for example, contour lines represent elevation on topographic maps), but a more natural way to think of them is in terms of continuous expanses, or surfaces.

The most common way to model a geographic surface is with a matrix of square cells, or pixels. Each cell represents a unit of area, such as a square meter, and stores a single piece of geographic information—typically, a measured or estimated value—at that location.

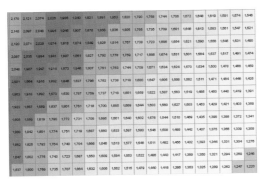

In this example, each cell represents thirty square meters of ground and stores a single representative elevation value.

This way of modeling surfaces is called the raster data model. It's commonly used for elevation and its derivations (slope, aspect); for temperature, precipitation, and land cover; for statistical data, such as densities and means; and especially for imagery.

The raster dataset is the basic storage unit for GIS data created according to the continuous surface view of the world. Raster datasets can be stored in geodatabases or in various standard image file formats, such as TIFF and JPEG.

Raster datasets in geodatabase format (left) and .tif format (right).

Feature classes and raster datasets are complementary. In many maps, raster datasets are used for background display, while feature classes are used for foreground display and analysis.

Figure 2-8

4) **Preview parcels.** First, we'll preview the parcels data, which comes from the City of Los Angeles.

Ⓐ In the Catalog Tree, under the City of LA folder, click *Parcels.shp* to highlight it (Figure 2-8).

The icon ⊞ signifies a polygon feature class in shapefile format. On the Contents tab, you see a thumbnail graphic of the data.

Ⓑ Click the Preview tab.

This gives you a larger and interactive view of the data. The previously disabled tools on the Geography toolbar (Figure 2-9) can now be used.

Figure 2-9

The Geography toolbar

Ⓒ On the Geography toolbar, click the Zoom In tool ⊕ .

Ⓓ Click and drag a box, like the one shown in Figure 2-10, anywhere on the data.

The size and location of your box doesn't have to match.

Figure 2-10

Figure 2-11

Ⓔ Zoom in farther until you can easily distinguish individual features, as in Figure 2-11.

This is the spatial data we need, showing individual parcel boundaries. By default, previewed features display with uniform symbology. Polygon features are yellow with gray outlines.

❻ On the Geography toolbar, click the Identify tool . Click any parcel to identify it. (Figure 2-12).

▷ The Parcels dataset is large, so the response may be a bit slow.

Hold the Shift or Control key to add features to the list.

Identify Results
Layers: \<Top-most layer\>

⊟ Parcels
　⊞ **195B137**
　⊞ 195B137

Location: (6414194.402880 1907011.465091)

Field	Value
FID	119331
Shape	Polygon
MAPSHEET	195B137
ID	443
MAP_REF	M B 654-31/32
TRACT	TR 21662
BLOCK	
LOT	72
PIN	195B137-443
AREA	8219.704838
PERIMETER	393.994398

Figure 2-12

Depending on exactly where you click, you may identify one or more features. On the left side of the Identify Results window, the identified parcel is listed by one of its attribute values. On the right side is the full set of attributes: mostly codes and identification numbers.

The vacancy attribute we need isn't here, but we do have an AREA attribute in unspecified units. Once we find out what the units are (which we'll do in Lesson 3), we can convert them to acres. Figure 2-13 provides some visual context for the size of a one-acre parcel compared to a 10-acre park.

Figure 2-13

OID	PIN	APN	ADDRESS	ZONECITY	ZONECODE	LANDUSE
0	151-5A151-996	4380015003		LA	RE15	010V
1	151-5A151-997	4380015003		LA	RE15	010V
2	151-5A151-1013	4380015003		LA	RE15	010V
3	151-5A173-1	5562006010		LA	RE40	010V
4	151-5A175-246	5567022011		LA	R1	010V
5	151-5A175-247	5567022018		LA	R1	010V
6	151-5A175-249	5556003026		LA	R1	010V
7	151-5A175-250	5556001001		LA	R1	010V
8	151-5A175-252	5556003028		LA	R1	010V
9	151-5A175-255	5567022002		LA	R1	010V
10	151-5A175-257	5569035008	1967 MOUNT OLY	LA	RE40	010V
11	151-5A175-258	5556002012		LA	R1	010V
12	151-5A175-259	5567022010		LA	R1	010V
13	151-5A175-260	5567022009		LA	R1	010V
14	151-5A175-262	5567022008		LA	R1	010V
15	151-5A175-264	5567022007		LA	R1	010V
16	151-5A175-265	5567028002		LA	R1	010V
17	151-5A175-266	5567022006		LA	R1	010V
18	151-5A175-274	5567022005		LA	R1	010V
19	151-5A175-275	5567022001		LA	R1	010V
20	151-5A175-276	5567022025		LA	R1	010V
21	151-5A175-277	5567022004		LA	R1	010V
22	151-5A175-279	5556005015		LA	R1	010V
23	151-5A175-281	5567022005		LA	R1	010V
24	151-5A175-282	5567022003		LA	R1	010V
25	151-5A175-291	5567022004		LA	R1	010V
26	151-5A175-293	5556003050		LA	R1	010V
27	151-5A175-294	5567022005		LA	R1	010V

(of 44462)

Preview: Table

Figure 2-14

G Close the Identify Results window.

5) Preview the table of vacant parcels. Since we don't have a vacancy attribute in the parcels shapefile, we'll look for it elsewhere.

A In the Catalog Tree, under the City of LA folder, click *Vacant Parcels.dbf.*

This is a stand-alone table. It has information about parcels, but no association with spatial features. Each of the 44,462 records represents a vacant parcel (Figure 2-14).

B Scroll across the attributes.

OID is a software-managed record identifier. PIN (Parcel Identification Number) and APN (Assessor Parcel Number) are user-managed identifiers. Several other attributes follow. In the LANDUSE field, all values end in "V," the code for vacancy.

C Click the Description tab.

The Description tab shows metadata, or data documentation. Complete metadata, which we'll look at later in the lesson, includes many elements. What you see here is called the Item Description, an overview of the dataset.

D Scroll down through the Item Description (Figure 2-15).

From the top, you see:

Figure 2-15

- The dataset name (Vacant Parcels) and file type (dBASE Table)
- A thumbnail image
- Tags that make the data searchable
- A summary of the intended use of the data
- A description of the data content
- Credits
- Use limitations
- Extent
- Scale Range

In this case, the summary contains a key piece of information. The records in this table can be joined, or matched, to features in the *Parcels* feature class using the PIN attribute, which is common to this table and the *Parcels* attribute table. This gives us a way to locate vacant parcels on the map. We'll get to this operation, called a table join, in Lesson 4.

6) **Preview cities.** Row 4 of the requirements table lists cities as needed spatial data. We have this data: we used it in Lesson 1, when we made a definition query on Los Angeles.

Ⓐ Click the Preview tab.

Ⓑ In the Catalog Tree, under ESRI.gdb, under Boundary, click the *City_ply* feature class to highlight it (Figure 2-16).

Figure 2-16

The preview changes from Table to Geography. The window is probably all yellow because you're still zoomed in to the area where you were previewing the *Parcels* feature class.

Ⓒ On the Geography toolbar, click the Full Extent button 🌐.

You zoom out to a view of all the features in the feature class.

Ⓓ In the Catalog Tree, still under the Boundary feature dataset, click *City_pt*.

The main difference between the two feature classes is that *City_pt* represents cities as points rather than polygons. (Hence the names *City_pt* and *City_ply*.) There's also a difference in spatial extent because *City_pt* includes a feature for Attu Station in Alaska, a place which is so far "west" that it is actually in the eastern hemisphere (Figure 2-17).

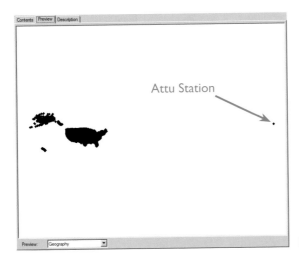

Figure 2-17

E Zoom in on the data to see individual features.

Why would the same features, such as these cities, be represented with two different types of shapes (points here and polygons in *City_ply*)? It's because they're appropriate for maps of different scale. For a national map, you would show cities as points. For a local map you might show them as polygons. For our requirement, which is to make sure that potential park sites lie inside the city limits, we need features that have boundaries. The polygon feature class is obviously the right one.

F Click the Full Extent button.

7)   **Preview the LA River.** In Row 5 of the requirements table, we have rivers as needed spatial data. In Lesson 1 we added the *River* feature class from the ESRI geodatabase. We also have another feature class to look at: *LARiver.shp* in the City of LA folder.

A Preview the geography of the *LARiver* shapefile.

Line features preview in dark blue. It looks like there's just one river in this feature class.

B At the bottom of the Preview tab, click the drop-down arrow and change the preview from Geography to Table.

C Scroll down to the bottom of the table.

In fact, the Los Angeles River *is* the only river in the feature class, but it's composed of 265 separate features (the FID, or feature ID, of the first record is 0). Why so many? As noted in Lesson 1 (see page 20), the answer has to do with attributes.

D Scroll across the table to see the attributes, then scroll up and notice the different values in the CAPACITY, DISCHARGE, and PROTECTION fields.

For our purposes, it doesn't really matter what these attributes mean. The point is that if the river were represented as a single feature, it would also have just one row in the attribute table. That would mean that only a single value could be stored for each attribute—fine for the river name (which doesn't change), but a problem for anything we might want to measure or describe at different locations along the river: flow, depth, water chemistry, navigability, or anything else. The creators of this data wanted to gather facts about the river at different places. To do that, they had to define the river as a spatially connected series of individual features.

We don't have that need. All we want is the spatial data. If we end up using this feature class in our analysis (rather than the *River* feature class in ESRI.gdb), we'll probably combine the 265 features into one.

8) Preview parks. In Row 6 of the requirements table, we need spatial data representing parks. We already know we have parks data: in Lesson 1, we symbolized and labeled the *Parkland* feature class. There's also a shapefile called *Parks* in the City of LA folder.

Ⓐ Set the preview back to Geography.

Ⓑ In the Catalog Tree, click *Parks.shp*.

Ⓒ Preview the geography, then the table (Figure 2-18).

Figure 2-18

It looks from the geography preview like the extent of the features corresponds to the city of Los Angeles. We can't be sure until we add the data to the map. The table preview shows us that there are 337 records (features) and just a few attributes.

Software-managed attributes

Every shapefile feature class has FID and Shape attributes that are created and managed by the software. The FID (feature identification) attribute stores a unique number for each feature. The Shape attribute stores the geometry type. Behind the scenes, it also links each feature to coordinates that define its spatial location. Measurement attributes, such as length and area, can be calculated for shapefiles, but if the values change—because of a spatial edit, for example—the software doesn't update them automatically.

A geodatabase feature class has up to four software-managed attributes. Like a shapefile, it has a feature identifier (called OBJECTID instead of FID) and a Shape attribute. The Shape_Length attribute stores the length of line and polygon features. This attribute doesn't exist for point features. The Shape_Area attribute stores the internal area of polygon features. It doesn't exist for point or line features. Shape_Length and Shape_Area are automatically kept up-to-date by the software.

Figure 2-19

Figure 2-20

D Set the preview back to Geography.

E In the Catalog Tree, under the ParkData folder, click *NewParks.shp*.

F Preview its geography and table.

This shapefile has just two features. One is Los Angeles State Historic Park and the other is Rio de Los Angeles State Recreation Area. Note the absence of length and area attributes. We can make them if we want—for example, the *Parks* shapefile has them—but they don't exist by default.

G Click the Description tab (Figure 2-19) and read the summary.

The data represents two newly developed parks. In Lesson 3, when we choose a parks feature class for the analysis, we'll have to make sure that it incorporates these two parks.

The ParkData folder also contains a raster dataset.

H Click the Preview tab.

I In the Catalog Tree, click *VistaHermosaPark.tif*.

The geography preview shows an aerial photograph of a park under development (Figure 2-20). Located just north of downtown, Vista Hermosa Park has since been completed. It's not included in the *NewParks* shapefile, however, so it's one more park to keep track of.

There's no table to preview. For most raster datasets, the only attribute to speak of is the value stored in each cell.

J On the Geography toolbar, click the Identify tool ⓘ.

K Identify a location anywhere on the image to see the RGB value of the identified pixel (Figure 2-21).

L Identify a few more pixels and close the Identify Results window.

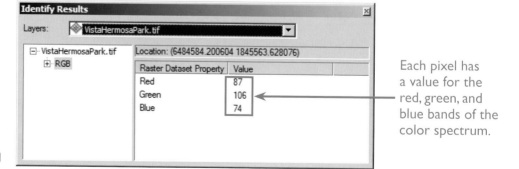

Figure 2-21

Each pixel has a value for the red, green, and blue bands of the color spectrum.

If you've taken a picture with a digital camera, you're probably familiar with the concept of raster data. A digital image captures a piece of the real world as a two-dimensional array of square cells, or pixels. Each pixel stores a number that corresponds to a specific intensity of light. A picture taken in sunlight is a composite of reflected light in different portions, or bands, of the visible electromagnetic spectrum: blue, green, and red. When these bands are displayed on your computer monitor, they blend to make a color that looks natural (Figure 2-22).

Images are a common form of raster data, but not the only one. Geographic phenomena with a continuous spatial distribution, like elevation and temperature, can also be represented by raster datasets, in which each cell stores a measured or estimated value.

A raster dataset composed of a single matrix of cell values is a single-band raster. One example is a digital elevation model (DEM), where each cell stores a single elevation value. Another example is a grayscale image. True color images, and most satellite images, are multiband, meaning that the raster dataset is composed of multiple spatially coincident cell matrices.

Figure 2-22

9) **Preview census units.** In Row 7 of the requirements table we decided to use census units as a proxy for neighborhoods.

Ⓐ In the Catalog Tree, under the census folder, click *tracts.shp* to preview its geography (Figure 2-23).

The data covers Los Angeles County. The tracts to the north are much bigger than the ones to the south. That's because census tracts are designed to have a fairly consistent population range, and the northern part of the county, with mountains and desert, is less densely populated.

Ⓑ Click the Zoom In tool 🔍 and zoom in somewhere on the southern part of the county.

Ⓒ In the Catalog Tree, click *block_groups.shp*.

Ⓓ Now click *block_centroids.shp*. If necessary, zoom in farther to see individual points.

Tracts are subdivided into block groups, and block groups are subdivided into blocks. A block centroid is a block represented spatially as a point rather than a polygon. (That's not so strange—we've seen the same thing with cities.) You can learn more about census units in the topic *Fundamentals of U.S. census geography*.

Ⓔ Click the Full Extent button 🌐.

Figure 2-23

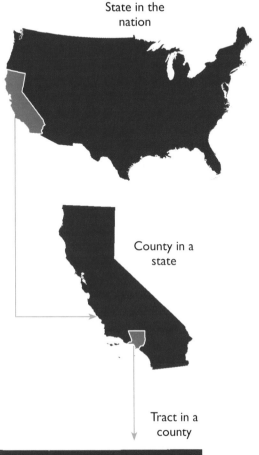

State in the
nation

County in a
state

Tract in a
county

The U.S. Census Bureau reports data by various geographic units. The top-to-bottom relationship shown here represents containment: the nation contains states, states contain counties, counties contain tracts, tracts contain block groups, and block groups contain blocks. Many other non-nesting reporting units are not shown.

Census tracts are relatively small subdivisions of a county. They typically have between 1,000 and 8,000 inhabitants and vary in size. They are designed to be fairly homogeneous with respect to demographic and economic conditions.

A block group is a cluster of blocks within a tract. A block group typically has between 600 and 3,000 inhabitants.

A census block (commonly an ordinary city block) is an area bounded by visible features, such as streets or railroad tracks, or by invisible boundaries, such as city limits. A block centroid is a census block represented as a point rather than a polygon. A centroid is located in the geographic center of the block it represents and has the attributes of that block.

The Census Bureau conducts a new census every ten years. The latest one was conducted in 2010. Professional demographers estimate values for the intervening years.

Block group in
a tract

Block in a block
group

Block centroid
in a block

Either block groups or tracts will satisfy our spatial data requirement for a neighborhood. Because of their point geometry, block centroids won't.

Thus far, we've confirmed that we have the spatial and attribute data listed in Rows 1–7 of the data requirements table. In three cases (rivers, parks, and census units), we'll have to choose between feature classes. We'll tackle that problem in Lesson 3. We still have more data requirements to consider and more data to preview. We also need to review our requirements for specificity. We'll do this in the next exercise.

F If you're going on to the next exercise, leave ArcCatalog and the data requirements table open; otherwise, from the ArcCatalog main menu, choose File → Exit. Save and close the data requirements table.

Exercise 2c: Reframe the problem statement

Some of the city council's guidelines were specific and measurable:

- On a vacant land parcel one acre or larger
- Within the LA city limits

Others were vague:

- As close as possible to the LA River. (Is there a maximum allowed distance from the river? If so, what is it?)
- Not in the vicinity of an existing park. (How close is "in the vicinity"?)
- In a densely populated neighborhood with lots of children. (How densely populated? How many children?)
- In a lower-income neighborhood. (How is "lower-income" defined?)
- Serving as many people as possible. (How big an area does a park "serve"?)

We can't do the analysis until we eliminate the vagueness.

1) Define proximity to the LA River. Unless we set a maximum distance limit, every vacant parcel in Los Angeles becomes a potential park candidate. That's absurd and could waste a lot of data processing time. We'll set three-quarters of a mile as an arbitrary outer limit. That stretches the idea of proximity somewhat, but it's just a cutoff point. Hopefully, we'll find some good locations that are closer than that.

A If necessary, start ArcCatalog. Under C:\UGIS\ParkSite, expand the SourceData folder and its subfolders.

B If necessary, use Windows Explorer to open the data requirements table from the MapsAndMore folder.

By default, you can't open this table from ArcCatalog, because ArcCatalog doesn't recognize the .doc file type. (It's not a geographic data format.) To add a file type, from the main menu, choose Customize → ArcCatalog Options. On the File Types tab, click New Type and enter the extension and description of the file type you want to recognize.

C In Row 5 of the data requirements table, in the DEFINED AS column, enter **<=0.75 miles** (Figure 2-24).

Figure 2-24

#	REQUIREMENT	DEFINED AS	SPATIAL DATA	ATTRIBUTE
1	land parcel		parcels	
2	vacant			land use
3	one or more acres			area
4	within city limits		cities	
5	near LA river	<= 0.75 miles	rivers	
6	away from parks		parks	
7	in a neighborhood		census unit	
8	densely populated			population de
9	lots of kids			age
10	low income			income

"<=" means "less than or equal to"

2) Define "not in the vicinity of a park." What minimum distance should a candidate site have to be from existing parks? In open-space planning, 0.25 miles is often used to define a convenient walking distance. (That's typically about a five-minute walk.) Following that standard, we'll say that a site is not in the vicinity of an existing park if the nearest park is at least 0.25 miles away.

A In Row 6 of the requirements table, in the DEFINED AS column, enter **>=0.25 miles**.

This measure is a simplification because it's based on straight-line distance. See Appendix A, *Analysis issues,* for further discussion.

Figure 2-25

TRACT	FIPS	POP2000	POP00_SQMI	POP2010	POP10_SQMI
101110	06037101110	4500	10227.8	4731	10752.3
101122	06037101122	3176	3113.7	3664	3592.2
101210	06037101210	6066	24264	5990	23960
101220	06037101220	3028	11214.8	3363	12455.6
101300	06037101300	3974	4096.9	4199	4328.9
101400	06037101400	3760	1579.8	3903	1639.9
102103	06037102103	1705	3552.1	1740	3625
102104	06037102104	2974	4248.6	3648	5211.4
102105	06037102105	1850	9736.8	1731	9110.5
102107	06037102107	3567	707.7	3851	764.1
103101	06037103101	2608	3180.5	2670	3256.1
103102	06037103102	4462	9295.8	4751	9897.9
103200	06037103200	5875	2005.1	6150	2099
103300	06037103300	3247	1364.3	3819	1604.6
103400	06037103400	6046	5927.5	6143	6022.5

Contents | Preview | Description

1 ▶ ▶| (of 2343)

Preview: Table

3) Define a "densely populated" neighborhood. As we make the rest of the requirements concrete, we'll also make sure that we have the appropriate data.

A In ArcCatalog, in the Catalog Tree, under the census folder, click *tracts.shp.*

B Preview its table and scroll across the attributes.

As shown in Figure 2-25, there are two population density attributes: POP00_SQMI and POP10_SQMI. These attributes store population per square mile for the census years 2000 and 2010, respectively.

C Preview the *block_groups.shp* table and scroll across its attributes.

The table has three attributes for total population: POP2000, POP2010, and TOTPOP_CY. (TOTPOP_CY, or total population for the current year, reflects 2011 projections by Esri demographers from the 2010 census data.) There is no population density attribute, but that turns out not to be a problem. Population density is total population divided by area. Since we have total population, we just need the area of each block group, which is something ArcMap can calculate automatically. So either feature class satisfies our need: *tracts* already has population density, and we can derive it from *block_groups*.

We still need a definition of "densely populated." To keep it simple, let's call a neighborhood densely populated if its value exceeds that of the city of Los Angeles as a whole. The population density of Los Angeles as of the year 2010 was 8,092.3 people per square mile. Rounding down to an even number, we'll use 8,000 people per square mile as our threshold.

D In Row 8 of the data requirements table, in the DEFINED AS column, enter `>=8,000 per sq mi`.

Threshold values

To find the population density of Los Angeles, as well as other threshold values for the analysis, we used online U.S. Census Bureau data, especially the QuickFacts page at quickfacts.census.gov.

4) **Define "lots of children."** Again, we'll look for attributes and then set a threshold.

A In ArcCatalog, scroll in the *block_groups* table and locate the POP18UP_CY attribute (Figure 2-26).

OWNER_CY	RENTER_CY	VACANT_CY	POP18UP_CY	MEDAGE_CY
177	32	5	647	34.4
16	672	62	1501	29.8
563	64	23	1708	43.1
207	126	16	710	34.5
308	746	56	2659	37.6
258	231	16	1470	30.9
593	34	30	1549	45.2
827	112	30	2915	31.8
138	381	15	1462	26.8
589	38	13	1860	44.4
167	813	47	2243	30.1
277	34	10	710	44.4
212	613	80	1277	36.2
115	279	32	1288	25.5
173	33	6	672	33.8
157	217	17	1264	30
0	0	0	0	0
37	284	46	764	27
166	68	7	625	36.9
53	954	200	2195	31.8

31 ▶ ▶| (of 6421)

Preview: Table

Figure 2-26

This is the population age 18 or older for the year 2011. If we define a child as a person under 18 (which is reasonable), then we can subtract the values in this field from those in TOTOP_CY to get the number of children.

ⓑ Preview the *tracts* table and scroll across its attributes.

Here, the population is segmented into age ranges (Figure 2-27). We can find the approximate number of children by adding the values in the AGE_UNDER5, AGE_5_9, AGE_10_14, and AGE_15_19 fields. This would actually be an overcount, because it would include some adults among the children, but basically we can obtain the required attribute from either table.

Figure 2-27

Adding values in these four fields gives the total number of children.

Because neighborhoods vary in size and population, we can make more valid comparisons if we base our threshold on a ratio rather than on an absolute number. In Los Angeles as a whole, 23.1 percent of the population is under 18. We'll therefore define a neighborhood as having "lots of children" if it meets or exceeds this value. (We can derive the percentage of children from our data with simple arithmetic.)

ⓒ In Row 9 of the requirements table, in the DEFINED AS column, enter **>=23%**.

ⓓ In the ATTRIBUTE DATA column, replace the entry "age" with **age under 18**.

5) **Define "lower income."** Let's preview our income attributes.

ⓐ In ArcCatalog, in the *tracts* table, scroll across the fields.

There don't appear to be any income attributes.

ⓑ Preview the *block_groups* table and scroll all the way to the right.

Metadata

Metadata is a description of what is known about a dataset. It serves two important purposes. First, it vouchsafes the integrity of data by explaining things such as how, when, and by whom the data was created. Second, it makes the data searchable. Metadata includes tags that identify essential properties of the data (for example, "rivers," "Los Angeles," and "2010") and other descriptions that make it possible to find specific datasets among large inventories of spatial data.

Metadata may be kept according to one of various official standards. Data created by government agencies, commercial data vendors, and many large enterprises typically conforms to one of these standards. Data created by small organizations or by individuals commonly does not. In ArcGIS, metadata can be displayed in a style that is suited to a particular standard. The default style is the Item Description, which displays a thumbnail image of the data and a small amount of important information. This style is suited to metadata that is not kept to an official standard. It can also be used to provide a filtered, summary view of metadata that is kept to an official standard. Anyone who creates and shares data should at least maintain metadata at the Item Description level.

To see the full metadata for a dataset that is kept to an official standard, you need to change the metadata style in ArcCatalog. All the styles, apart from Item Description, are similar, and all afford access to the full set of metadata—no matter what standard it conforms to—though they may present the information slightly differently.

In ArcCatalog, full metadata appears under the heading ArcGIS Metadata. This heading is found immediately beneath the Item Description (which displays with all styles). In special cases, particularly with older metadata or metadata created outside the ArcGIS system, you may need to look under the heading FGDC Metadata (read-only), which follows the ArcGIS metadata, in order to see all the available information.

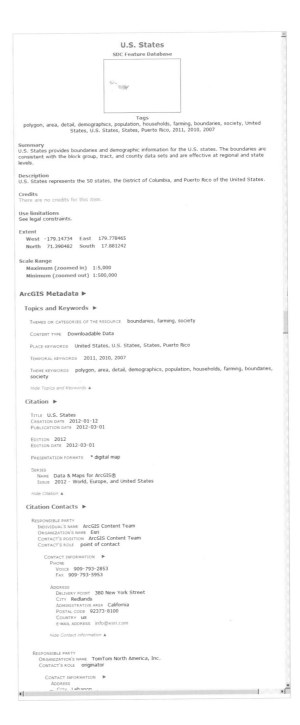

The last four attributes might be income measures, but we can't be sure without decoding the field names. To do this, we'll look at the metadata—not the item descriptions we looked at before, but the full data documentation.

C From the main menu, choose Customize → ArcCatalog Options. In the dialog box, click the Metadata tab.

D Change the metadata style from Item Description to North American Profile of ISO19115 2003 (Figure 2-28). Click OK.

Figure 2-28

Changing the metadata style gives you access to the full set of metadata for the dataset. See the topic *Metadata* for more information.

E Click the Description tab.

F Scroll down past the Item Description until you see the ArcGIS Metadata heading.

The ArcGIS metadata is divided into sections that can be expanded or collapsed. (They are expanded by default.)

G Click the Topics and Keywords heading to collapse it (Figure 2-29).

H Collapse the next several headings (Citation, Citation Contacts, and so on) until you come to the Fields heading.

This heading contains the metadata we're interested in.

I Scroll through the Fields metadata.

For each field, there are several pieces of information. Asterisked information (*) is created and maintained by ArcGIS, and has to do with technical properties of the field. Nonasterisked information is created by a person.

J Scroll down to the last field, PCI_CY, and read the field description.

Click a heading to collapse or expand it.

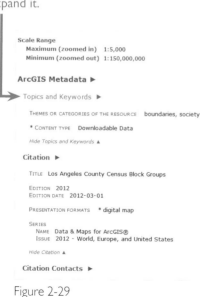

Figure 2-29

The field is per capita income for the current year. As with the TOTPOP_CY attribute described earlier, this is a forecast based on 2010 census data.

ⓚ Read the descriptions for the three fields above PCI_CY.

The attributes are aggregate (total) household income, average household income, and median household income (Figure 2-30).

We don't want to use aggregate income because it's not standardized: block groups with more people are likely to have higher aggregate incomes. The other three are good possibilities. Median income is a statistical midpoint: it marks the value that half the households are above and half are below. We'll adopt this measure on the grounds that it's less sensitive to extreme values. (A millionaire in a low-income neighborhood might significantly change the average income, but not the median income.)

The median household income for Los Angeles for the years 2007–2011 is $50,028. Rounding off, we'll call a neighborhood "lower income" if the median household income is $50,000 or less.

ⓛ In Row 10 of the requirements table, in the DEFINED AS column, enter <=$50,000.

ⓜ In the ATTRIBUTE DATA column, replace "income" with **median hh income**.

Figure 2-30

6) **Define "serving the most people."** Finally, we want to know which potential site serves the most people. Anyone can come to a park, so for this criterion we want to count all the people nearby, regardless of their demographic profile. The attribute we need for this is total population, which we have in both the *tracts* and *block_groups* feature classes.

We'll treat this guideline as a preference. If half a dozen sites meet our other requirements, we prefer those serving more people overall to those serving fewer. Eventually, this preference may have to be subjectively weighed against others. For example, which is better: a park closer to the river that serves fewer people, or a park farther from the river that serves more people? (How much closer? How many more people?)

To define the size of the area served by a park, we'll apply the standard of easy walking distance discussed earlier, and say that a park serves anyone who lives within a quarter mile of it. "Serving the most people" therefore means having the largest population within a 0.25 mile radius.

ⓐ In ArcCatalog, preview the *block_centroids* table and scroll across its attributes.

This table also has a total population attribute (POP2010). We can't use block centroids as our spatial data for neighborhoods—we need polygons—but we can conveniently use them to sum population. Given the boundary of a park service area, ArcMap can count the block points that fall within it and add their population values.

❸ In Row 11 of the requirements table, in the DEFINED AS column, enter `<=0.25 miles`.

Your requirements table should look like Figure 2-31.

#	REQUIREMENT	DEFINED AS	SPATIAL DATA	ATTRIBUTE DATA	DATA SET	PREPARATION
1	land parcel		parcels			
2	vacant			land use		
3	one or more acres			area		
4	within city limits		cities			
5	near LA river	<= 0.75 miles	rivers			
6	away from parks	> 0.25 miles	parks			
7	in a neighborhood		census unit			
8	densely populated	>= 8,000 per sq mi		population density		
9	lots of kids	>= 23%		age under 18		
10	low income	<= $50,000		median hh income		
11	serving the most people	<= 0.25 miles		population		
12	final map		political boundaries			
13	final map		roads			
14	final map		relief			
15	final map		imagery			

Figure 2-31

❸ From the ArcCatalog main menu, choose Customize → ArcCatalog Options.

❹ On the Metadata tab, change the style back to Item Description. Click OK.

❺ From the ArcCatalog main menu, choose File → Exit.

❻ Save and close the data requirements table.

We can now state the problem in measurable terms that allow us to solve it with GIS tools. Someone might take issue with our interpretation of the guidelines, but that's fine—we'll always be happy to improve our methodology. For now, we can state our project analysis requirements as follows:

We want to locate a site for a new park

- On a land parcel
- That is vacant
- That is one or more acres in size
- That is within the LA city limits
- That is within 0.75 miles of the LA River (preferring closer sites)
- That is more than 0.25 miles from the nearest park
- That is in a census unit where the population density is 8,000 or more people per square mile
- And where at least 23 percent of the population is under 18 years old
- And where median household income is $50,000 or less
- And where, given that all other conditions are satisfied, the total population within a 0.25-mile radius is maximized.

We've formalized the data requirements in a table and confirmed that the essential data is available. In some cases, there are multiple datasets that contain suitable features and attributes. In the next lesson, we'll compare these datasets and choose which ones to use in the analysis.

Lesson

3 Choose the data

CHOOSING THE DATASETS TO USE

in the analysis is our first goal in this lesson. We came across a few situations in Lesson 2 (with rivers, parks, and census units) in which different datasets stored similar features and attributes. How do we choose one dataset over another? Here are some considerations:

- One may represent features with a more suitable geometry type for your needs (think of cities as polygons versus cities as points).
- Among datasets of the same geometry type, one may represent features with a more appropriate degree of detail.
- One dataset may be of more recent vintage than another. Or it may be more complete, including features that the other lacks.
- One dataset may be more accurate than another when viewed against imagery or other basemaps. Or it may have been created by a more authoritative source. Or it may conform better to another dataset you've already decided to use.
- One dataset may need less preliminary processing of spatial or attribute data to make it ready for analysis.

Our second goal is to choose a coordinate system for the analysis. In reality, that decision is often determined by prevailing standards in your organization, or by more or less default best choices for a given study area. For example, it makes sense for us to use the State Plane Coordinate System of 1983 because it is specifically designed to minimize spatial distortion for more than a hundred local areas (like ours) within the United States. While the choice of a coordinate system bears directly on our project, Exercise 3b also contains background material about coordinate systems (what they are, why they matter, how they're managed) that is important to know for any GIS work.

Lesson Three roadmap

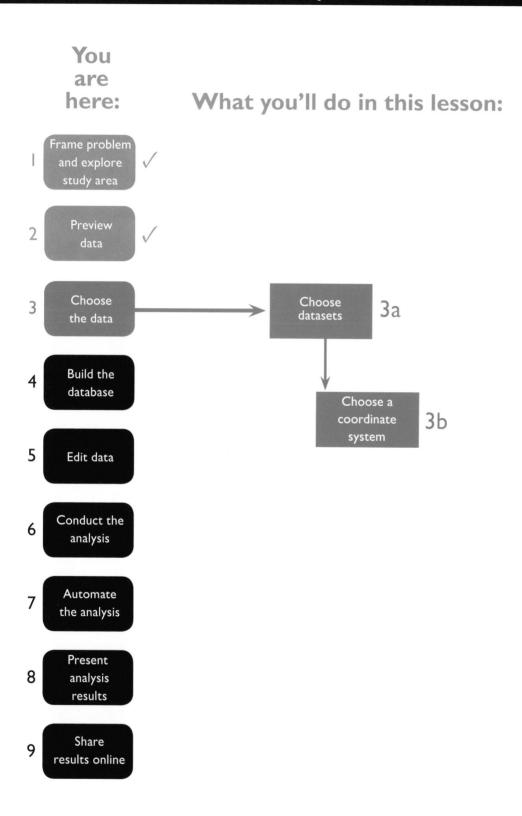

You are here:

What you'll do in this lesson:

1 Frame problem and explore study area ✓

2 Preview data ✓

3 Choose the data

3a Choose datasets

3b Choose a coordinate system

4 Build the database

5 Edit data

6 Conduct the analysis

7 Automate the analysis

8 Present analysis results

9 Share results online

Exercise 3a: Choose datasets

In this exercise, we'll add layers to ArcMap, compare them, and decide which datasets to use in the analysis. We'll add this information to the data requirements table we used in the last lesson.

1) Start ArcMap. We'll work in ArcMap in this lesson and access data from the Catalog window.

🅐 Click the Windows Start button, and choose All Programs → ArcGIS → ArcMap 10.1.

▷ If you have a shortcut on your desktop, you can double-click the ArcMap 10.1 icon 🔍 instead.

🅑 In the Getting Started dialog box, under New Maps, click My Templates to highlight it (Figure 3-1). Click Blank Map and click OK.

Figure 3-1

🅒 Make sure the Catalog window is either open or visible as a tab on the right side of the ArcMap window.

▷ If necessary, on the Standard toolbar, click the Catalog window button 🗄.

2) Choose the parcel data. Parcels are the individual property boundaries from which our candidate park sites will eventually be selected.

🅐 In the Catalog window, click the plus sign next to Folder Connections to expand it.

🅑 Expand C:\UGIS.

🅒 Expand the ParkSite, SourceData, and City of LA folders.

In Lesson 2, we previewed the *Parcels* shapefile and the *Vacant Parcels* table in the City of LA folder. We don't have any other parcel data, so there's no real choice to make here.

    D Using Windows Explorer, open the data requirements table from C:\UGIS\ParkSite\MapsAndMore.

        ▷ If you didn't complete Lesson 2, you can download the lesson results from the Understanding GIS web site.

    E In Row 1 of the table, in the DATASET column, enter **Parcels**.

    F In Row 2, under DATASET, enter **Vacant Parcels**.

We need to join the *Vacant Parcels* table to the *Parcels* attribute table to see vacant parcels on the map. We'll do this table join—along with all our other data preparation—in Lesson 4, but we'll make a note of it here.

    G In Row 2 of the requirements table, in the PREPARATION column, enter **join table to Parcels**.

We also need to calculate parcel acreage to satisfy our third requirement.

    H In Row 3, under DATASET, enter **Parcels**. Under PREPARATION, enter **calculate area** (Figure 3-2).

Figure 3-2

#	REQUIREMENT	DEFINED AS	SPATIAL DATA	ATTRIBUTE DATA	DATASET	PREPARATION
1	land parcel		parcels		Parcels	
2	vacant			land use	Vacant Parcels	join table to Parcels
3	one or more acres			area	Parcels	calculate area
4	within city limits		cities			
5	near LA river	<= 0.75 miles	rivers			
6	away from parks	> 0.25 miles	parks			
7	in a neighborhood		census unit			

3) Choose the city limits data. The requirement in Row 4 is that the new park be within the city limits. We assumed this meant that we'd need spatial data representing the boundary of Los Angeles (which we have), but actually, there's an even simpler solution.

    A In ArcMap, in the Catalog window, under City of LA, right-click *Vacant Parcels.dbf* and choose Item Description.

The Item Description window opens. Its Description and Preview tabs give you access to the same information you got from ArcCatalog in the last lesson.

    B In the Item Description window, click the Preview tab.

C Scroll across the attributes and make sure you can see the ZONECITY field (Figure 3-3).

Figure 3-3

This field tells you which city each vacant parcel is recorded in.

D Make sure you're scrolled to the top of the table. Right-click the ZONECITY field heading and choose Sort Ascending.

The first record is in LA.

E Right-click the ZONECITY field heading again and choose Sort Descending.

The table doesn't change, which tells you that every record in the field has the same value. This means that all our vacant parcels—and consequently, all our potential park sites—are prequalified as belonging to Los Angeles. It turns out that we don't need a spatial boundary after all to guarantee that a parcel lies within the city limits.

You can resize the window by dragging any side or corner.

F Close the Item Description window.

G In Row 4 of the requirements table, under DATASET, enter **Vacant Parcels**.

H Also in Row 4, under SPATIAL DATA, delete "cities." Under ATTRIBUTE DATA, enter **city name**.

4) Add and symbolize the river data. To analyze the distance of parcels to the river, we need spatial data representing the Los Angeles River. We can use either the *River* feature class in the ESRI geodatabase or the *LARiver* shapefile. We'll compare these two datasets against an imagery basemap.

A In ArcMap, on the Standard toolbar, click the Add Data drop-down arrow 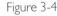 and choose Add Basemap.

B In the Add Basemap dialog box, click Imagery and click Add.

By default, the *Basemap* layer is zoomed to the extent of the whole world.

C In the Catalog window, expand ESRI.gdb and expand Hydro.

D Drag and drop *River* onto the map display.

E In the Table of Contents, right-click the *River* layer and choose Zoom To Layer (Figure 3-4).

Figure 3-4

Layers usually draw in randomly selected colors. You may not have noticed, but this layer has drawn in exactly the same shade of blue both times you've added it to ArcMap. Let's see why.

❻ In the Table of Contents, click the symbol patch for the *River* layer to open the Symbol Selector (Figure 3-5).

When a feature class name matches a symbol name, that symbol is assigned automatically.

Figure 3-5

In the Symbol Selector, notice the River symbol in the scrolling box. Because the feature class name (*River*) matches the name of a predefined symbol, the layer is symbolized automatically.

❼ Close the Symbol Selector.

Symbols and styles

ArcGIS has thousands of predefined symbols for map features, organized into "styles" on a mostly thematic basis: the Caves style, the Crime Analysis style, and so on. By default, symbols from two styles are loaded in the Symbol Selector. One is the ESRI style, which contains common map symbols; the other is your personal style (which is empty unless you add symbols to it). The set of available symbols changes according to whether you're symbolizing point, line, or polygon features.

To load additional styles, click the Style References button on the Symbol Selector and check the styles you want. Symbols from these styles will be added at the bottom of the Symbol Selector. You can also search for symbols by typing a keyword in the Search box at the top of the Symbol Selector.

Styles contain not only feature symbols, but also labels, color ramps, north arrows, scale bars, and many other map elements. To view or edit the contents of a style, choose Style Manager from the Customize menu.

The only river we want to see on the map is the LA River. In Lesson 1, we solved this problem with a definition query. Another approach is to symbolize the Los Angeles River and no other features.

H Open the layer properties for the *River* layer and click the Symbology tab.

I In the Show box, click Categories.

You symbolize by category when the attribute of interest is a name, a description, or a number that isn't a quantity, such as a code or ranking. The default Unique Values method assigns a different symbol to each unique value in the specified field.

J Make sure the Value Field drop-down list is set to NAME. Click the Add Values button (Figure 3-6).

Figure 3-6

Click here to symbolize a descriptive attribute.

Click here to pick values from the NAME field.

Figure 3-7

K In the Add Values dialog box, click Complete List.

L Scroll down to "Los Angeles River." (The list is alphabetical.) Click on it to highlight it (Figure 3-7) and click OK.

On the Symbology tab, "Los Angeles River" is added and assigned a symbol from the current color ramp. By default, other features in the layer will get the symbol next to <all other values>. We don't want to see those features, however.

M Uncheck the <all other values> check box.

N Right-click the symbol patch next to Los Angeles River and choose Properties for Selected Symbol(s) to open the Symbol Selector.

O Change the color to Apatite Blue (Figure 3-8). Change the line width to 2. Click OK on the Symbol Selector.

Figure 3-8

P Compare your settings to Figure 3-9 and click OK on the Layer Properties dialog box.

Uncheck this box ———→

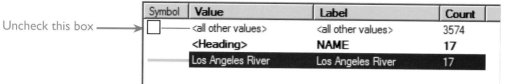

Symbol	Value	Label	Count
☐	<all other values>	<all other values>	3574
	<Heading>	**NAME**	**17**
	Los Angeles River	Los Angeles River	17

Figure 3-9

On the map, only the Los Angeles River is symbolized (Figure 3-10).

Q In the Catalog window, under City of LA, drag and drop *LARiver.shp* onto the map.

You should get a Geographic Coordinate Systems Warning (Figure 3-11). If the meaning of this warning is unclear to you—as it probably will be unless you know how ArcGIS manages coordinate systems—be patient until we get to Exercise 3b.

Figure 3-10

Figure 3-11

Figure 3-12

R Click Close on the Geographic Coordinate Systems Warning dialog box.

The *LARiver* layer is added to the map.

S In the Table of Contents, click the *LARiver* symbol patch to open the Symbol Selector.

T Change the color to Cretan Blue (Figure 3-12) and the width to 2. Click OK.

5) Choose the river data. Even at this relatively small (zoomed out) scale, it's obvious that the two representations of the river have different spatial extents. One goes to the sea while the other stops abruptly halfway there. Let's find out why.

Ⓐ In the Catalog window, expand the MapsAndMore folder.

Ⓑ Drag and drop *Los Angeles.lyr* on the map.

▷ If you didn't create this file in Lesson 1, you can download the Lesson 1 results from the Understanding GIS web site.

Ⓒ In the Table of Contents, right-click the *Los Angeles* layer and choose Zoom To Layer (Figure 3-13).

This is the layer file (the file of saved layer properties) that you made in Lesson 1.

Ⓓ Open the layer properties for the *Los Angeles* layer.

Ⓔ On the Layer Properties dialog box, click the Symbology tab, and then the Definition Query tab.

As reflected on the map, the symbol is a hollow yellow outline and there is a definition query on Los Angeles.

Figure 3-13

Ⓕ Click the Source tab to see the Data Source panel (Figure 3-14).

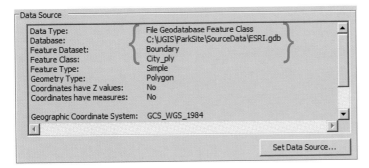

Figure 3-14

A layer file isn't a feature class—it doesn't store feature coordinates and attributes. Every layer in a map document must point to a feature class on disk, and the same goes for a layer file. This layer file points to the *City_ply* feature class.

Ⓖ Close the Layer Properties dialog box.

We can see the likely explanation for the different extents of the two river layers: in the *LARiver* layer, the river stops at the city boundary. The City of Los Angeles, which provided the source data for this layer, doesn't need to maintain features beyond its own jurisdiction. (It's interesting that even a natural feature such as a river might be defined in part by administrative or political considerations.)

Ⓗ Zoom in 🔍 on the source (the northwestern end) of the river. Set the map scale to 1:24,000.

Ⓘ Pan 🖑 along the river and visually compare the two representations. Go all the way to where the *LARiver* layer stops at the city boundary.

The two representations of the river aren't identical, but they're pretty close. At very large scales, like 1:2,500, the river in *LARiver* is noticeably more sinuous and conforms more closely to the imagery, but for our analysis, either dataset is perfectly adequate. We'll use the *LARiver* dataset, which is already clipped to our area of interest.

Ⓙ In Row 5 of the table, under DATASET, enter **LARiver**.

When we previewed this dataset in Lesson 2, we saw that it was composed of 265 features. We can combine these into a single feature with a data processing operation called Dissolve. That will help get the data ready for Lesson 6, where we'll need to create a 3/4-mile buffer (proximity zone) around the river.

Ⓚ In Row 5, under PREPARATION, enter **dissolve** (Figure 3-15).

Figure 3-15

#	REQUIREMENT	DEFINED AS	SPATIAL DATA	ATTRIBUTE DATA	DATASET	PREPARATION
1	land parcel		parcels		Parcels	
2	vacant			land use	Vacant Parcels	join table to Parcels
3	one or more acres			area	Parcels	calculate area
4	within city limits			city name	Vacant Parcels	
5	near LA river	<= 0.75 miles	rivers		LARiver	dissolve
6	away from parks	> 0.25 miles	parks			
7	in a neighborhood		census unit			
8	densely populated			population density		

6) Save the map document. This is a good time to save the map document. Let's take a minute to understand how ArcMap encourages us to organize maps and data.

Ⓐ In ArcMap, open the Catalog window.

Ⓑ At the top of the Catalog window, expand the Home – Documents\ArcGIS folder (Figure 3-16).

Default Home folder ⟶

Figure 3-16

Every map document has a home folder, which is simply the folder in which it is saved. In a new ArcMap session, before a map document has been saved, the designated home folder is MyDocuments\ArcGIS.

○ On the Standard toolbar, click the Save button 🖫 .

○ Navigate, if necessary, to the MapsAndMore folder. Name the map document **Lesson3a** and click Save.

In the Catalog window, the home folder changes to ParkSite\MapsAndMore (Figure 3-17). The name of the open map document is boldfaced.

After saving

Figure 3-17

Home folders and default geodatabases

The home folder helps you keep map documents, datasets, and related files in the same place. This keeps the path relationship between layers and their data sources simple and easy to manage. It also simplifies the creation of map packages, which are bundles of maps and data that you can share (see Lesson 9). Most nondataset files that you save from a map document, such as layer files, graphs, and bookmarks, are directed to the home folder.

In addition to a home folder, every map document has a default geodatabase. This is the location to which data processing outputs are directed by default. Unless otherwise specified, the default geodatabase (which is actually named Default.gdb) is located in MyDocuments\ArcGIS. You can choose a different default geodatabase when you start a new map document. You can also change the default geodatabase during a session by right-clicking the geodatabase you want in the Catalog window and choosing Make Default Geodatabase.

7) **Add and symbolize the park data.** To analyze the location of candidate sites with respect to existing parks, we need spatial data representing parks. We'll compare the *Parkland* dataset we used in Lesson 1 to the *Parks* dataset we previewed in Lesson 2.

○ In the Table of Contents, turn off the *River* layer and click its minus sign to collapse it.

▷ Leave the *LARiver* layer turned on.

○ In the Catalog window, under ESRI.gdb, expand Landmark.

○ Drag and drop *Parkland* onto the map.

○ From the City of LA folder, drag and drop *Parks.shp* onto the map.

▷ Close the Geographic Coordinate Systems Warning if you get it.

Figure 3-18

Figure 3-19

Figure 3-21

The order of layers in the Table of Contents should look like Figure 3-18. Your symbology for the parks layers will be different.

  E In the Table of Contents, right-click the symbol patch for the *Parks* layer to open the color palette. Change the fill color to Chrysoprase (Figure 3-19).

  F Right-click the *Parkland* symbol patch and change its fill color to Tzavorite Green.

  G Open the layer properties for the *Parks* layer and click the Display tab.

  H In the Transparent % box, type **50**. Click OK.

The symbology makes it clear where the two layers agree and disagree on the representation of features (Figure 3-20).

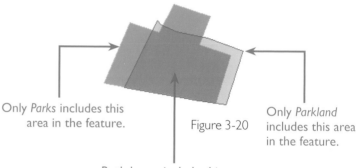

Only *Parks* includes this area in the feature.

Figure 3-20

Only *Parkland* includes this area in the feature.

Both layers include this area.

8) **Examine a park in two layers.** Let's start with a close look at an example park. Your map should still be zoomed in to where the river crosses the city limits. The scale should still be 1:24,000.

  A On the Tools toolbar, click the Measure tool. At the top of the dialog box, make sure the Measure Line tool ~ is selected.

  B Click the Choose Units button ▾, point to Distance, and choose Miles, if necessary (Figure 3-21).

  C Move the Measure dialog box away from the river, if necessary.

We're going to follow the river north about two and three-quarters miles.

  D Click at the end of the river to start the measurement. Move the mouse pointer north along the river (you don't have to drag).

As you move the pointer the Length value changes in the Measure dialog box.

  E When you reach the top of the map display—assuming your measurement is still less than 2.75 miles—press and hold the letter C on your keyboard.

The mouse pointer changes to the Pan icon .

F Pan north and release the C key to continue the measurement.

G When the length of the measurement reaches about 2.75 miles, double-click to end the measurement.

Half a mile due east of the end of your measurement is a park.

H Identify ⓘ the park.

Its name is Pecan Playground (Figure 3-22). The attributes in the Identify window come from the *Parks* layer, not from *Parkland,* because of the layer order in the Table of Contents.

Figure 3-23

Figure 3-22

By default, the topmost layer is identified. You can choose other layers from this drop-down list.

I In the top pane of the Identify window, right-click the name Pecan Playground. On the context menu, choose Zoom To.

The map zooms in on the park (Figure 3-23).

J Close the Identify window and the Measure dialog box.

K Turn the *Parkland* layer off and on a few times.

Neither feature conforms to the image: both encroach, for example, on the surrounding streets. The *Parks* feature seems better, however, because it includes the swimming pool and the play area at the north end of the block. We'll come back to this place in Lesson 5 to do some spatial editing.

Figure 3-24

Data Frame Tools toolbar

We noted in Lesson 1 that a tool called My Places is a good way to store and manage locations independently of particular map documents. (Within a map document, it's easier to use spatial bookmarks.) To use My Places, we need to add a toolbar.

L From the ArcMap main menu, choose Customize →Toolbars → Data Frame Tools (Figure 3-24).

M On the toolbar, click the My Places button ⚑.

N On the right side of the My Places dialog box, click Add From and choose Current Extent.

An entry called My Place is added to the box (Figure 3-25).

← Click here

Figure 3-25

O Right-click the highlighted entry and choose Rename. Replace the default name with **Pecan Playground** and press Enter.

In Lesson 1, we saved our spatial bookmarks to an external file. We'll load these into My Places now.

P In the My Places dialog box, click Load. Navigate, if necessary, to the MapsAndMore folder.

Q Click Lesson 1 Places.dat to highlight it and click Open.

▷ If you don't have this file, you can download the Lesson 1 results from the Understanding GIS web site.

Your My Places dialog box should look like Figure 3-26.

Figure 3-26

R In the My Places dialog box, click Dodger Stadium to highlight it and click Zoom To.

Your view is now centered on Dodger Stadium.

S Close the My Places dialog box.

9) **Choose the park data.** We'll follow the river upstream and look at some more parks to help us reach a decision.

A In the Table of Contents, drag the *Los Angeles* layer directly underneath the *LARiver* layer.

B Set your map scale to **1:36,000**. Start panning northwest along the river's course.

You'll soon come to Griffith Park, the vast park at the river's elbow. Both the river and the city limits run along the park's edge. Notice that to the east (Glendale) and north (Burbank) of the city limits, there are features from the *Parkland* layer, but no corresponding *Parks* features (Figure 3-27). This is another case of jurisdiction. The City of Los Angeles, which supplied the *Parks* dataset, doesn't maintain data for parks in other cities.

Figure 3-27

That raises a question. The new park has to be a quarter mile away from existing parks: does that mean any park in any city or just any park in Los Angeles? We'll interpret it to mean any park in any city, since by and large, there are no residency requirements for park use. On that basis, we should choose the *Parkland* feature class for our analysis. We don't know if it's more or less accurate on the whole (we only examined one park), but it covers a part of our area of interest that the *Parks* feature class doesn't.

C In Row 6 of the requirements table, under DATASET, enter **Parkland**.

Figure 3-28

10) Add more parks data. In Lesson 2, we previewed the *NewParks* feature class and the *VistaHermosaPark* raster dataset. We should find out if these features are included in the *Parkland* layer.

Ⓐ In ArcMap, in the Catalog window, expand the ParkData folder. Add *NewParks.shp* to the map.

▷ Close the Geographic Coordinate Systems Warning if you get it.

Ⓑ In the Table of Contents, drag the *NewParks* layer directly above the *Parks* layer.

Ⓒ Right-click the symbol patch for the *NewParks* layer and click Macaw Green (Figure 3-28).

Ⓓ Zoom to the *NewParks* layer (Figure 3-29).

▷ How? Right-click the layer in the Table of Contents and choose Zoom To Layer.

Figure 3-29

Ⓔ Turn the *NewParks* layer off and on a couple of times.

You can see parks on the basemap imagery, but they're not represented in the *Parkland* layer. Another data preparation task in Lesson 4 will be to load these two park features into the *Parkland* feature class.

Ⓕ In the Catalog window, under ParkData, add *VistaHermosaPark.tif* to the map.

▷ Close the Geographic Coordinate Systems Warning if you get it.

Ⓖ In the Table of Contents, collapse the layer by clicking its minus sign.

Ⓗ Zoom to the *VistaHermosaPark.tif* layer.

There's no *Parkland* feature here, either, or we'd be looking at it. In Lesson 5, we'll create a park feature in this location.

I Turn the *VistaHermosaPark.tif* layer off and on a couple of times (Figure 3-30).

The basemap imagery, in which the park looks more developed, is probably newer.

J Open My Places 🏳 . Click Add From and choose Current Extent.

K Right-click the new entry and choose Rename. Name it **Vista Hermosa Park**. Click Close on the My Places dialog box.

L In Row 6 of the requirements table, under PREPARATION, enter **load data and edit features**.

11) Choose the census data. In Lesson 2, we decided that we would represent neighborhoods with census units. Our two choices are block groups or tracts. Block groups meet all our attribute needs, either directly or indirectly. Because they are smaller than tracts, block groups also portray demographic patterns in more detail.

A In Rows 7 through 10 of the requirements table, under DATASET, enter **block_groups**.

Population density and age under 18 attributes don't exist as such in the *block_groups* table, but we can derive them.

B In Rows 8 and 9, under PREPARATION, enter **calculate**.

The last analytical requirement is to count how many people live within 0.25 miles of each proposed park site. In Lesson 6, we'll do this by drawing quarter-mile rings (buffers) around the proposed sites, counting the census block points in each ring, and summing their populations.

C In Row 11, under DATASET, enter **block_centroids** (Figure 3-31).

Figure 3-30

The *VistaHermosaPark.tif* image was downloaded from DataDoors, a geographic image data service, at http://esri.datadoors.net.

Figure 3-31

#	REQUIREMENT	DEFINED AS	SPATIAL DATA	ATTRIBUTE DATA	DATASET	PREPARATION
1	land parcel		parcels		Parcels	
2	vacant			land use	Vacant Parcels	join table to Parcels
3	one or more acres			area	Parcels	calculate area
4	within city limits			city name	Vacant Parcels	
5	near LA river	<= 0.75 miles	rivers		LARiver	dissolve
6	away from parks	> 0.25 miles	parks		Parkland	load data and edit features
7	in a neighborhood		census unit		block_groups	
8	densely populated	>= 8,000 per sq mi		population density	block_groups	calculate
9	lots of kids	>= 25%		age under 18	block_groups	calculate
10	low income	<= $50,000		median hh income	block_groups	
11	serving the most people	<= 0.25 miles		population	block_centroids	
12	final map		political boundaries			
13	final map		roads			
14	final map		relief			

12) Choose the final map data. The last rows of the table list data that we may need for cartographic reasons in the final map.

Row 12 lists political boundaries. Our source data has feature classes of cities, counties, and states. Until we design our final map, we won't know for sure, but we foresee a likely need for city and county boundaries, and not for state boundaries.

ⓐ In ArcMap, in the Catalog window, under ESRI.gdb, expand Boundary.

In Lesson 2, we previewed *City_ply* and saw that its extent covered the entire United States. That's more data than we'll need. We never previewed the *County* feature class, so let's do that now.

ⓑ Under Boundary, right-click the *County* feature class and choose Item Description.

In the Item Description window, on the Description tab, the thumbnail image indicates that the data covers the United States. Thumbnail images aren't automatically synchronized with the data, though, and may offer an outdated snapshot. It's a good idea to preview the geography, which is always current.

ⓒ In the Item Description window, click the Preview tab (Figure 3-32). Leave the Item Description window open.

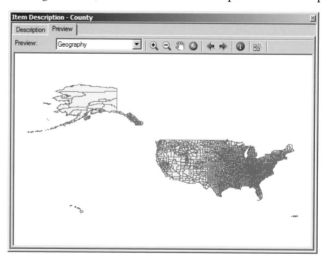

Figure 3-32

The preview confirms the spatial extent of the feature class.

ⓓ In Row 12 of the requirements table, under DATASET, enter **City_ply**. Press Enter to start a new line and enter **County**.

ⓔ In Row 12, under PREPARATION, enter **reduce extent** next to *City_ply* and *County*.

In Row 13, we have a need for roads listed.

F In ArcMap, in the Catalog window, under ESRI.gdb, expand Transport.

G Click the *Mjr_rd* feature class.

The Item Description window updates to preview the *Mjr_rd* geography.

H Click the Description tab, and read the summary, then close the Item Description window.

The data covers the greater Los Angeles area, which is appropriate.

I In Row 13 of the requirements table, under DATASET, enter **Mjr_rd**.

Relief and imagery are available as online basemap layers.

J In Rows 14 and 15 of the requirements table, under DATASET, enter **basemap** (Figure 3-33).

#	REQUIREMENT	DEFINED AS	SPATIAL DATA	ATTRIBUTE DATA	DATASET	PREPARATION
1	land parcel		parcels		Parcels	
2	vacant			land use	Vacant Parcels	join table to Parcels
3	one or more acres			area	Parcels	calculate area
4	within city limits			city name	Vacant Parcels	
5	near LA river	<= 0.75 miles	rivers		LARiver	dissolve
6	away from parks	> 0.25 miles	parks		Parkland	load data and edit features
7	in a neighborhood		census unit		block_groups	
8	densely populated	>= 8,000 per sq mi		population density	block_groups	calculate
9	lots of kids	>= 25%		age under 18	block_groups	calculate
10	low income	<= $50,000		median hh income	block_groups	
11	serving the most people	<= 0.25 miles		population	block_centroids	
12	final map		political boundaries		City_ply County	reduce extent reduce extent
13	final map		roads		Mjr_rd	
14	final map		relief		basemap	
15	final map		imagery		basemap	

Figure 3-33

13) **Save your work.** All the datasets we need for the analysis have been specified, so we can save and close our work.

A Save and close the data requirements table.

B In ArcMap, zoom to the *Los Angeles* layer.

C Save the map document.

D Optionally, remove the Data Frame Tools toolbar.

▷ How? From the ArcMap main menu, choose Customize → Toolbars and uncheck Data Frame Tools.

E If you are continuing to the next exercise now, leave ArcMap open; otherwise, exit ArcMap.

In the next exercise, we'll take a step back from the project. Analysis operations take place within a particular coordinate system, and choosing that system is an important part of setting up the project. To make a good choice, you should have some background knowledge of what coordinate systems are and how they're managed in ArcMap.

Exercise 3b: Choose a coordinate system

A coordinate system is a mesh of perpendicular intersecting lines superimposed on a surface. The point of intersection of any two lines is a unique location, which can be specified with two values (a coordinate pair). The values are measurements from a given reference point in a given unit of measure. The unit of measure may be an angular unit, like degrees, or a length unit, like feet or meters.

Coordinate systems can be applied to any surface, but we're interested in the surface of the earth. If we represent the earth with a spherical model, like a globe, we have a curved surface to deal with. If we represent it with a flat model, like a map, the surface is flat.

We apply a geographic coordinate system to round or roundish earth models. Geographic coordinate systems use an angular unit of measure because angles are better for measuring locations on a curved surface. We apply a projected coordinate system to flat earth models. Projected coordinate systems use a length unit of measure. Figure 3-34 illustrates both types.

Every usable spatial dataset has a coordinate system. The locations of its features are specified by coordinates that are correct within its own framework, but which would be wrong or absurd in another system. Fifty degrees of arc isn't the same as 50 meters, which isn't the same as 50 feet. Knowing the coordinates of a point doesn't tell you where the point is on the earth unless you also know what the coordinate system is.

There are many different geographic and projected coordinate systems. As long as two datasets use the same system, their features will align correctly on a map. Datasets with different

Figure 3-34

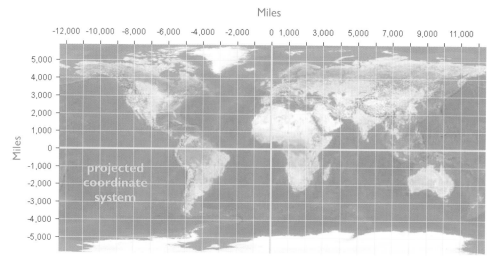

systems need to be reconciled. This basically means that one dataset's coordinate system has to be mathematically converted, or projected, to the other. This can be done with data processing tools. It is also done automatically by ArcMap when you add layers to a map document. What ArcMap reconciles, however, are the coordinate systems of the layers in the map, *not* the coordinate systems of the datasets that the layers point to. This reconciliation of coordinate systems in a map document is called "on-the-fly" projection.

We'll see how ArcMap manages on-the-fly projection of coordinate systems. We'll also look at how map display and map measurements change with different geographic and projected coordinate systems. Finally, we'll select a coordinate system for our analysis project.

1) **Check the coordinate system of a dataset.** Every spatial dataset has a coordinate system. Let's see how to find out what it is.

Ⓐ If necessary, start ArcMap and open a blank map.

  ▷ If ArcMap is already running, on the Standard toolbar, click the New button ☐ and open a blank map.

Ⓑ In the Catalog window, under ESRI.gdb, expand Boundary. Right-click the *State* feature class and choose Properties.

Ⓒ In the Feature Class Properties dialog box, click the XY Coordinate System tab.

The bottom panel of the tab displays the current coordinate system (Figure 3-35).

"GCS" stands for Geographic Coordinate System. "WGS_1984" is the World Geodetic System of 1984, also called WGS84.

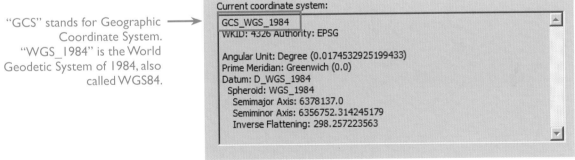

Figure 3-35

The name of the coordinate system is GCS_WGS_1984. Under the name are details giving the essential properties of the system:

- Its unit of measure (degrees of angle)
- Its prime meridian (Greenwich)
- Its datum (D_WGS_1984)

The datum includes the name and dimensions of a particular spheroid, or roundish earth model.

WGS_1984 is a geographic coordinate system. Geographic coordinate systems, and datasets that use them, are said to be "unprojected." If we checked the other feature classes in ESRI.gdb, we'd find that they all have this same geographic coordinate system.

D Close the Feature Class Properties dialog box.

2) Add data to ArcMap. Let's add this unprojected data to ArcMap and see what it looks like.

A Add *State* as a layer to ArcMap (Figure 3-36).

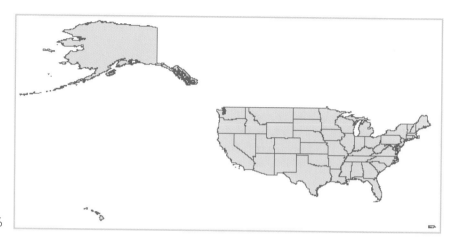

Figure 3-36

B Open the layer properties for the *State* layer and click the Source tab.

The same coordinate system information (minus the details) that you saw in the feature class properties can be found in the layer properties as well (Figure 3-37).

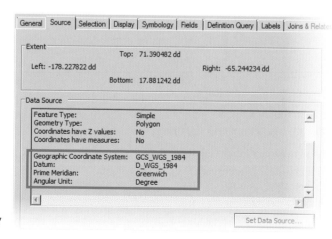

Figure 3-37

C Close the Layer Properties dialog box.

D Hover over the map (Figure 3-38).

Figure 3-38

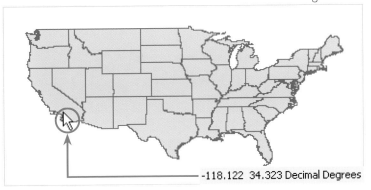

-118.122 34.323 Decimal Degrees

In the lower right corner of the application window, the coordinates of the mouse pointer are reported in decimal degrees. The first coordinate (longitude) tells you the angular position east or west of the prime meridian. The second coordinate (latitude) tells you the angular position north or south of the equator.

Negative values are used for west longitude (as here) and south latitude.

Degrees Minutes Seconds and Decimal Degrees

Angle measurements are commonly expressed in degrees, minutes, and seconds (DMS), where a circle has 360 degrees, a minute is 1/60th of a degree, and a second is 1/60th of a minute. ArcMap can express latitude-longitude values this way, but it makes calculations in decimal degrees (DD), which is a conversion of DMS to decimal notation. The following two expressions are equivalent:

-118.122, 34.323 (DD)

-118° 7' 19", 34° 19' 23" (DMS)

E From the ArcMap main menu, choose View → Data Frame Properties.

F In the Data Frame Properties dialog box, click the Coordinate System tab (Figure 3-39).

A data frame ▤ is a container for layers in a map document. Every map document starts with one data frame, named Layers, which is the top entry in the Table of Contents. You can insert more data frames to manage separate sets of data—essentially, multiple maps—within your map document. Why would you want to do that? It's usually a layout consideration: you may want to print a single map sheet that contains a main map, an overview map, an inset map, or some other combination of views. (We'll see how data frames work in map layouts in Lesson 8.)

One crucial function of a data frame is to enforce the spatial alignment of the layers it contains. An empty data frame, in a new map document, isn't associated with any coordinate system. When you first add a layer to it, it adopts that layer's coordinate system as a standard. Any layers added thereafter are converted (projected) by ArcMap into that same system. This process is called on-the-fly projection and happens automatically as new layers are added.

G Close the Data Frame Properties dialog box.

Figure 3-39

3) Project data on the fly. To see how on-the-fly projection works, we'll insert a second data frame, add a layer that has a different coordinate system from the *State* layer, then copy the *State* layer into the new data frame. To see the contents of both data frames at the same time, we have to open another viewer window.

Ⓐ On the Tools toolbar, click the Create Viewer Window tool ⊞ and drag a box around the data (Figure 3-40).

Figure 3-40

Figure 3-41

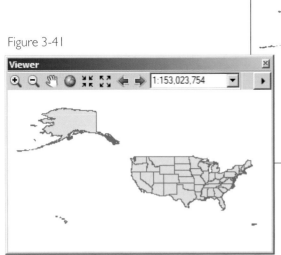

Move the viewer window by dragging its title bar. Resize it by dragging a side or corner.

When you release the mouse button, you'll have a viewer window (Figure 3-41) that you can move and resize however you like. Viewer windows are usually used to give you multiple vantage points on a map, but in this case, the purpose is to preserve a snapshot of the data as it looks in a geographic coordinate system.

At the top of the viewer window is a toolbar with navigation tools. Viewer windows can be zoomed and panned independently of the main map.

Ⓑ Experiment with panning and zooming in the viewer window. When you're finished, go back to the full extent ⊙.

Ⓒ From the main menu, choose Insert → Data Frame.

A new data frame, named "New Data Frame," is added to the Table of Contents. Its name is boldfaced to indicate that it's the active data frame, the one we're working with. (The title bar of the viewer window tells you that the data frame named "Layers" is inactive.) The main map window is empty because the new data frame doesn't contain any layers.

Ⓓ From the main menu, choose View → Data Frame Properties.

  ▷ A shortcut is to double-click the data frame name in the Table of Contents.

Ⓔ If necessary, click the Coordinate System tab.

The current coordinate system is "No coordinate system." We just wanted to confirm that a new data frame isn't associated with any coordinate system.

F Close the Data Frame Properties dialog box.

G Add the Imagery basemap (Figure 3-42).

▷ How? Click the Add Data drop-down arrow and choose Add Basemap. In the dialog box, click Imagery.

H Hover over the map.

The coordinates, in the lower right corner of the ArcMap window, are in meters. That's a sign that you're working with a projected coordinate system, one based on length measurements rather than angles.

Figure 3-42

I Open the data frame properties for New Data Frame again.

The data frame's coordinate system, adopted from the *Basemap* layer, is now WGS_1984_Web_Mercator_Auxiliary_Sphere (Figure 3-43). A projected coordinate system essentially consists of a map projection (in this case, a form of the Mercator projection) applied to a geographic coordinate system (in this case, GCS_WGS_1984). See the topics *Geographic coordinate systems* and *Projected coordinate systems* for more information.

WGS_1984_Web_Mercator_Auxiliary_Sphere
WKID: 3857 Authority: EPSG

Projection: Mercator_Auxiliary_Sphere
False_Easting: 0.0
False_Northing: 0.0
Central_Meridian: 0.0
Standard_Parallel_1: 0.0
Auxiliary_Sphere_Type: 0.0
Linear Unit: Meter (1.0)

Figure 3-43

J Close the Data Frame Properties dialog box.

K In the Table of Contents, under the inactive Layers data frame, right-click the *State* layer and choose Copy.

L Right-click on the name New Data Frame and choose Paste Layer(s).

M Under New Data Frame, right-click the *State* layer and choose Zoom To Layer.

The *State* layer aligns with the *Basemap* layer (Figure 3-44). In the process, its spatial properties have changed. Compare its appearance in the main map window to the viewer window. Notice especially the shape of Alaska.

N Close the viewer window.

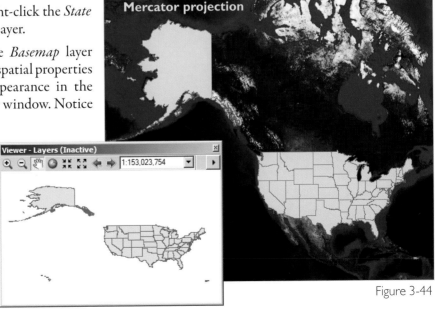

Figure 3-44

4) Make an area measurement. Coordinate systems change measurements as well as appearances. The Mercator projection we're using now is a conformal projection, which means that it shows shapes correctly. (Look at Alaska on a globe and you'll see this is true.) On the other hand, it distorts area measurements—pretty significantly in extreme latitudes.

Ⓐ Zoom in on Alaska. (It doesn't matter how far exactly.)

Ⓑ On the Tools toolbar, click the Measure tool 🔧.

Ⓒ At the top of the Measure dialog box, click the Clear and Reset Results button ✕ and click the Measure A Feature button + .

Ⓓ Click the Choose Units button ▾ , point to Area, and choose Miles.

Ⓔ Click anywhere inside Alaska (Figure 3-45). Leave the Measure dialog box open.

Figure 3-45

The area measurement for Alaska is 3,141,292 square miles.

Ⓕ Under New Data Frame, open the *State* layer attribute table.

Ⓖ Select the record for Alaska (Figure 3-46). (It should be the second one.) Scroll all the way to the right to see the SQMI attribute.

Click the gray square to select the record.

NO_FARMS07	AVG_SIZE07	CROP_ACR07	AVG_SALE07	SQMI	Shape_Len
48753	185	3142958	90.57	51649.05	22.59
686	1285	86238	83.12	581369.17	834.25
15637	1670	1205425	206.85	113997.77	23.90
49346	281	8432221	152.17	53178.77	22.04
81033	313	9464647	418.16	158144.83	57.60
37054	853	11483936	163.58	104093.95	22.02
4916	83	163686	112.19	4962.77	13.31
2546	200	432773	425.39	1975.78	8.88
-99	-99	-99	-99	62.19	1.1

Figure 3-46

The value in the table, which is correct, is 581,369 square miles. (You can find other values cited in reference sources, but they're similar to this one.) That means that the map measurement is an exaggeration by a factor of more than five.

Figure 3-47

❶ At the top of the table window, click the Clear Selection button ▣ . Close the table.

The data frame's current coordinate system

The Measure tool is doing its job correctly: it has measured the area of Alaska as distorted by the Mercator projection. (The Mercator projection has its virtues, but making area measurements in high latitudes is not one of them.)

5) Change the data frame's coordinate system. You can change the data frame's coordinate system whenever you want—before or after adding layers.

Ⓐ Open the data frame properties for New Data Frame.

On the Coordinate System tab, the current coordinate system is selected in the upper scrolling box (Figure 3-47).

Ⓑ Scroll to the top of the box (Figure 3-48).

Coordinate systems are organized in folders. The two main folders are Geographic Coordinate Systems and Projected Coordinate Systems. Both contain several subfolders. The Web Mercator Auxiliary Sphere coordinate system is in the Projected Coordinate Systems folder. Within this folder, it is located in the World folder, because it is a projected coordinate system designed to show the whole world at once.

Figure 3-48

The more you learn about coordinate systems, the more useful this folder structure becomes. At the outset, it can be a bit overwhelming.

Ⓒ Under Projected Coordinate Systems, expand the Continental folder, and then the North America folder.

The first coordinate system in this folder should be Alaska Albers Equal Area Conic. This system, unlike the Mercator, represents area (the size of features) correctly.

Ⓓ Click Alaska Albers Equal Area Conic to select it (Figure 3-49) and click OK.

▷ If you get a warning, click Yes to use the system anyway.

Figure 3-49

Geographic coordinate systems

A geographic coordinate system (GCS) is based on a spheroidal model of the earth. Sometimes a perfect sphere is used, but usually a slightly squashed "oblate spheroid" is used to reflect the fact that the earth bulges at the equator and is flattened at the poles.

Sphere   Oblate spheroid
                (much exaggerated)

The reference lines of a geographic coordinate system are parallels and meridians. Parallels are lines that circle the globe parallel to the equator. Meridians are lines perpendicular to the equator that converge at the poles. By convention, the origin of the system (its 0,0 coordinate), is the intersection of the equator and the prime meridian, the meridian passing through Greenwich, England.

Geographic coordinates, commonly called latitude-longitude values, are measurements of angle, not distance. Angles are a constant unit of measure on a sphere, while distances are not (because meridians converge).

Angle measurements are usually expressed in degrees, minutes, and seconds. A degree has 60 minutes; a minute has 60 seconds.

Latitude is angular position north or south of the equator. The equator is 0° latitude, the North Pole is 90° north, and the South Pole is 90° south.

Longitude is angular position east or west of the prime meridian. The prime meridian is 0° longitude. Its anti-meridian (on the other side of the world) is both 180° east and 180° west.

A latitude-longitude pair defines a unique position on the earth's surface.

The unique location of Dodger Stadium, for example, would be written like this: **34°4'26"N, 118°14'27"W** and spoken like this:

"34 degrees, 4 minutes, 26 seconds north latitude; 118 degrees, 14 minutes, 27 seconds west longitude."

For computer calculations, these values are converted to decimals. The location of Dodger Stadium in "decimal degrees" is 34.073, -118.24. The minus sign is used for west longitude and south latitude.

The fact that there are many different geographic coordinate systems is a source of trouble for GIS users. What makes two systems different is disagreement about the exact latitude-longitude values of particular locations. Why is there disagreement about that? In simple terms, it's because different spheroid models of the earth have been developed over time by different earth scientists using different technologies. Changing the shape or size of the model ends up changing the coordinates of points on the surface—usually not very much, but sometimes, in sensitive applications, enough to be of concern. This issue is taken up in more detail in the *Datums* topic later in this lesson.

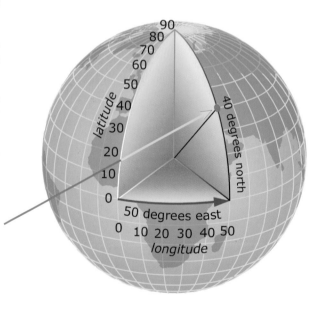

Projected coordinate systems

The earth's surface can be modeled very well on a spheroid, but not as well (except over small areas) on a plane. To make a map, you more or less have to flatten a sphere, which is like squaring a circle, only harder. It can't be done without radically adjusting the spatial properties and relationships of features on the surface: their shapes, sizes, and relative distances and directions.

The name for any such radical adjustment is a map projection. A projection is a mathematical formula (there are lots of different ones) for translating the world into flat space. All map projections introduce spatial distortion. They are variously designed to minimize certain kinds of distortion or to distribute it in certain ways over the map surface. Some projections correctly preserve feature shapes but distort their areas. Some preserve areas but distort shapes. Some compromise. Some have special properties, such as keeping true distance measurements from a single point to all others, or ensuring that courses of constant compass bearing are plotted as straight lines. The smaller the area being mapped, the less distortion there is of any spatial properties. Areas up to medium-sized countries (say about the size of Nigeria or Bolivia) can be mapped with distortion so low as to be insignificant for most purposes.

A projected coordinate system consists of a map projection, a length-based unit of measure, an origin point for measurements, and other parameters, such as standard lines that define the distortion pattern on the map. Manipulating these parameters is what allows you to customize a coordinate system for a specific area of interest. Because a projection is applied to a particular spheroid and its definition of latitude-longitude values, a projected coordinate system also includes, or presupposes, a geographic coordinate system.

The idea of projection includes both going from a geographic (unprojected) system to a projected system, and going from one projected system to another (sometimes called reprojection). To go from one projected system to another, ArcMap undoes the map projection, goes back to the underlying geographic coordinate system, and applies a new projection to it. ArcMap stores thousands of map projection formulas and can run these calculations quickly.

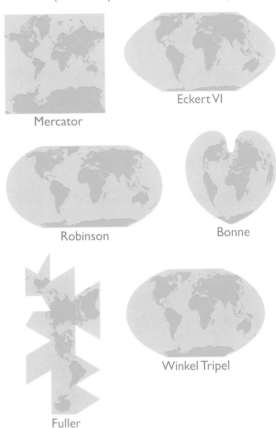

Some common world map projections
(and a couple of uncommon ones)

Mercator

Eckert VI

Robinson

Bonne

Fuller

Winkel Tripel

When you add an unprojected dataset to a map (that is, a dataset that stores feature coordinates as latitude-longitude values), the data still has to be projected in some sense in order to be viewed as a flat map on your monitor. In ArcMap, this default "pseudoprojection" has the display properties of a map projection (specifically, the Plate Carree), but none of the other properties or parameters of a projected coordinate system.

Geographic pseudoprojection

The new coordinate system is applied to the data frame. Both the *Basemap* layer and the *State* layer are projected (reprojected, if you like) on the fly. Your map should look like Figure 3-50. (You may have to click the Refresh button in the lower left corner of the map window to redraw the map.)

Figure 3-50

Albers Equal Area projection for Alaska

Choosing a coordinate system

A suitable coordinate system for your project depends largely on your area of interest, and may depend also on whether you need to preserve a particular spatial property, such as shape or area, without distortion. A few systems are common standards. For example, web-based maps typically use the Web Mercator coordinate system. Local-area maps within the United States are frequently based on the State Plane system, while local-area maps around the world (including the United States) often use the Universal Transverse Mercator (UTM) system.

On the Data Frame Properties Coordinate System tab, you can set a spatial filter 🔽 so that only coordinate systems covering your area of interest are displayed. (This doesn't mean all the results are appropriate for your map. Systems designed for world maps are displayed with every spatial filter, but may not be a good choice for a local map.) You can also enter text ⌷California▾⌷ to find coordinate systems that include this text as part of their name. This is often a good way to find systems specifically suited to your area of interest.

6) **Measure Alaska again.** Equal-area projections (of which the Albers Equal Area Conic is one) preserve the spatial property of area, or size. The trade-off is that they don't represent shapes correctly. Because this projection is customized for Alaska, however, most distortion of *all* types is pushed outside the area of interest.

Ⓐ Click on the Measure dialog box to make it active.

▷ Make sure the Measure A Feature tool is still selected.

Ⓑ Click on Alaska.

The area is given accurately as 581,332 square miles (Figure 3-51).

Ⓒ Close the Measure dialog box.

Ⓓ Zoom to the full extent (Figure 3-52).

Figure 3-51

Figure 3-52

This coordinate system is obviously unsuitable for a world map. It doesn't show the whole world for one thing. The price for mapping Alaska with very low distortion is that places such as Australia and South America are severely distorted—which is fine, as long as we keep our map zoomed in to Alaska.

Let's summarize how things stand with the layers in this data frame:

- The geographic coordinate system of the *State* layer is GCS_WGS_1984. It has no projected coordinate system.
- The projected system of the *Basemap* layer is Web_Mercator. Its underlying geographic system is GCS_WGS_1984.
- The data frame has been set to the Alaska Albers Equal Area Conic projected system. Its underlying geographic system is GCS_North_American_1983.
- The *State* and *Basemap* layers have been projected on the fly into the Alaska Albers system.

No matter how often we change coordinate systems, ArcMap keeps the data in alignment.

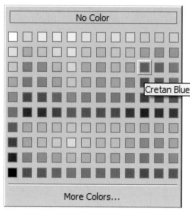

Figure 3-53

7) **Add another layer.** Let's get back to our true area of interest.

🅐 In the Catalog window, from the City of LA folder, drag and drop *LARiver* onto the map.

🅑 Zoom to the *LARiver* layer and turn off the *State* layer.

🅒 Change the color of the *LARiver* symbol to Cretan Blue (Figure 3-53). Change its line width to 2.

A coordinate system meant for Alaska (or the world) isn't appropriate for Southern California. Local mapping needs across the United States are served by a system called the State Plane Coordinate System. This isn't a single coordinate system for the entire country, but a patchwork of systems, each of which covers a state or part of a state. Together, they assure that for whichever part of the country you want to map, you get uniformly low distortion. California is divided into six State Plane zones, with Los Angeles falling in Zone 5 (Figure 3-54).

The State Plane Coordinate System divides states into zones and applies a locally optimized coordinate system to each zone. Vertical zones are based on the Transverse Mercator projection. Horizontal zones are based on the Lambert Conformal Conic projection.

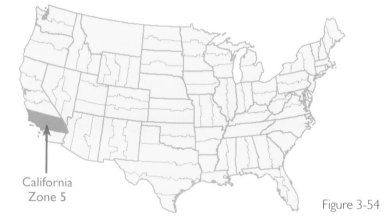

California Zone 5

Figure 3-54

8) **Change the data frame's coordinate system.** We'll change the data frame's coordinate system again, this time to State Plane California Zone 5.

🅐 Open the data frame properties for New Data Frame.

🅑 On the Coordinate System tab, scroll to the bottom of the list and expand the Layers folder.

Within this folder are the three native coordinate systems of the layers in the data frame.

🅒 Expand the coordinate system NAD_1983_StatePlane_California_V_FIPS_0405_Feet.

This is the coordinate system used by the *LARiver* layer. (Strictly speaking, it is the coordinate system of the *LARiver* shapefile.)

D Click the layer name to highlight it
(Figure 3-55).

This resets the current coordinate system of the data
frame. The State Plane California Zone 5 system is
a projected coordinate system based on a Lambert
Conformal Conic map projection.

E Scroll to the bottom of the lower box showing the
details of the current coordinate system
(Figure 3-56).

Figure 3-55

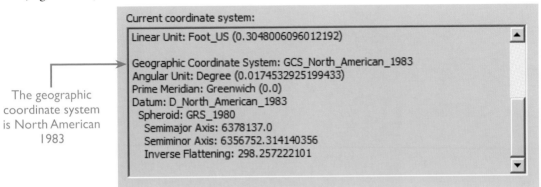

The geographic
coordinate system
is North American
1983

Figure 3-56

Its underlying geographic coordinate system is North American 1983.
This is different from the WGS 1984 system used by both the *State* and
Basemap layers.

F Click OK on the Data Frame Properties dialog box.

A warning appears (Figure 3-57). It's a variation of the
geographic coordinate systems warning we've seen before.
Both warnings alert you to a mismatch between the
geographic coordinate system of the data frame and one or
more of its layers. (You get the warning shown here when
you change the data frame's coordinate system; you get the
other warning when you add a nonmatching layer.)

Why does ArcMap keep issuing these warnings, and how
serious are they? Perfect projection or reprojection of data
into a different coordinate system requires that the "from"
and "to" systems have the same geographic coordinate
system. If they don't, it means they have a basic disagreement about
the latitude-longitude values of locations. Projection can't fix that
disagreement. Projection converts a given set of angular coordinates to
planar coordinates. It doesn't convert one set of angular coordinates to
another.

Figure 3-57

In this exercise, and elsewhere in the book, we've been mixing layers
based on two geographic coordinate systems: WGS 1984 and North
American 1983. Whenever we do so, ArcMap lets us know that it can't

guarantee perfect alignment of the data. The disagreements between these two systems are very small (they rarely translate into more than a few feet on the ground), but ArcMap doesn't know what we're doing with the data or how much accuracy we need. There is a way to reconcile different geographic coordinate systems—it's called a geographic transformation or a datum transformation—but we'll postpone doing one until the next lesson.

Ⓖ Click Yes on the warning.

Your map should look like Figure 3-58. The State Plane California Zone 5 coordinate system is the one we'll use for our analysis.

Figure 3-58

Ⓗ Save the map document as **Lesson3b** in the MapsAndMore folder.

Ⓘ Exit ArcMap.

Geographic coordinate systems warnings

When you get a geographic coordinate systems warning, check boxes allow you to turn them off, either for the duration of the ArcMap session or for all future sessions. Unless you're an expert, you shouldn't check "Don't warn me again ever." If you do so by mistake, you can reinstate the warning by opening AdvancedArcMapSettings.exe in the ArcGIS\Desktop10.1\Utilities folder. In the Advanced ArcMap Settings Utility, click the TOC/Data tab and uncheck the box "Skip datum check."

In the next lesson, we'll resume work on the project by taking care of the preparation tasks we listed in the data requirements table.

Datums

A geographic coordinate system is defined by three things: an angular unit of measure (usually degrees), a prime meridian (usually Greenwich), and a datum. The datum is the part that gives people trouble. To understand it, let's start with the shape of the earth.

The earth isn't a perfect sphere, or even a mathematically regular spheroid. It's a lump with an uneven shape owing to different concentrations of mass (and therefore unequal gravity) over its surface. In addition, it has topographic features such as mountains and valleys.

When the spatial positions of features are determined—as was formerly done by survey, and is now mostly done by satellite—they are first determined on the earth's surface. These raw measurement values are then mathematically "leveled" to a geoid. A geoid is the (still gravitationally lumpy) shape that the earth would have if it were covered by a mean sea level surface—in other words, if it had no topography.

The shape of the geoid, however, is too complex to be a working model. So the next step is to move the measurements from the geoid to a spheroid: a model with a regular, non lumpy shape.

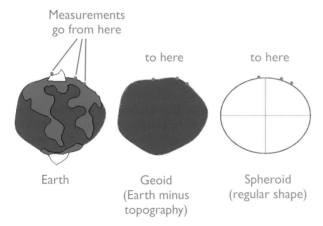

Measurements go from here

to here

to here

Earth

Geoid
(Earth minus topography)

Spheroid
(regular shape)

That's where the datum comes in. The datum is two things: first, it's a chosen spheroid, which could be WGS 1984, GRS 1980, Clarke 1866, Bessel 1841, or a number of others. (The world is standardizing on the GRS 1980 spheroid, but isn't all the way there yet.) Second, it's a mathematical orientation, or "fit," of the geoid to the spheroid. In the transfer of measurements from ge-

oid to spheroid, some error will be introduced because the lumps have to be smoothed out. How that error is distributed is the "fit." One approach is to make the fit really good for one part of the world, such as North America, and not to worry about the rest. That's a local datum. It's designed to maintain high accuracy for measurements over a limited area. The other approach is to average the error over the whole surface. That's an earth-centered datum. It's designed to maintain high accuracy for the world as a whole.

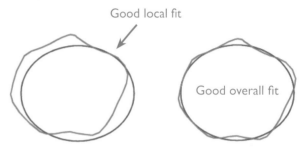

Good local fit

Good overall fit

When two geographic coordinate systems are different, it's usually because the datums are different (which, in turn, is either because the spheroids are different or the fit is different). When you get a coordinate systems warning in ArcMap, one possibility is to ignore it and leave the data slightly out of alignment. Depending on your needs for accuracy, this may be an entirely sensible choice. The amount of misalignment depends on the datums involved and the part of the world being mapped, but in the mismatch North Americans usually encounter (between the World Geodetic System of 1984 and the North American Datum of 1983), it typically doesn't exceed a few feet. At most scales, this isn't noticeable.

The other option is to reconcile the systems through a geographic transformation. Transformations are often done in conjunction with a coordinate system projection. Like projections, they can be permanently applied to datasets with data processing tools, or they can be done on the fly in ArcMap. They require some expert knowledge. There are default methods to convert one spheroid to another, but there aren't default fits, because the right fit depends on your area of interest. The document geographic_transformations.pdf, in the ArcGIS\Desktop10.1\Documentation folder, can help you find the right fit for an area of interest.

Lesson

4 Build the database

WE'VE COME TO A TURNING POINT

in our project. In the early, exploratory phases, we looked at data—both in and out of maps—and evaluated it from various standpoints. We worked a lot with layer properties, but we didn't operate directly on the data. We'll do that now.

We have three main tasks in this lesson. First, we'll create a geodatabase to store the data for the project. That raises the question whether the project really needs its own database. Couldn't we continue to use the collection of folders and files we've been working with so far? Yes, we could, but there are good reasons to make a new database:

- We can keep the project data in a single location, rather than having it distributed among various workspaces.
- We can keep the project data separate from datasets that we don't need.
- We can streamline the project data by eliminating unnecessary features and attributes, while maintaining the original source data as a backup.
- We can impose uniform standards like a common data format and consistent names for datasets.
- It will be easier to share the project data with others and to reuse it ourselves.

Our second task will be to populate the new geodatabase with the datasets that we chose in the last lesson. This will be our introduction to the geographic data processing or "geoprocessing" tools that make ArcGIS for Desktop so powerful.

Our third task will be to complete the preparation tasks in the data requirements table. Besides using geoprocessing tools, we'll do spatial queries, table calculations, and other operations.

Lesson Four roadmap

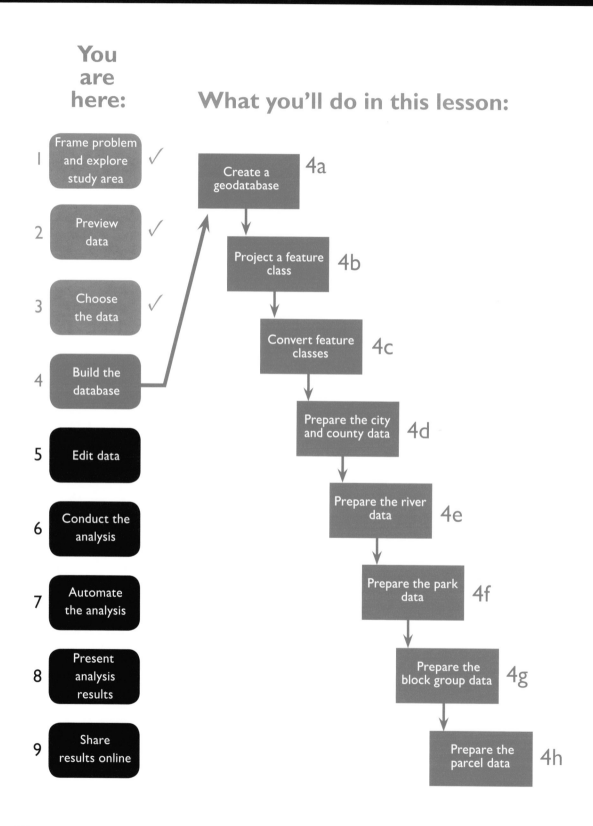

You are here:

1. Frame problem and explore study area ✓
2. Preview data ✓
3. Choose the data ✓
4. Build the database
5. Edit data
6. Conduct the analysis
7. Automate the analysis
8. Present analysis results
9. Share results online

What you'll do in this lesson:

4a Create a geodatabase

4b Project a feature class

4c Convert feature classes

4d Prepare the city and county data

4e Prepare the river data

4f Prepare the park data

4g Prepare the block group data

4h Prepare the parcel data

Exercise 4a: Create a geodatabase

In this short exercise, we'll create the database for our project. Our database is a simple one, but it still requires some thought about data format, organization of inputs and outputs, naming conventions, and other things. Our considerations are laid out in the topic *Project database considerations*.

1) Create a geodatabase. We'll work in ArcMap in this lesson, but we'll use the Catalog window a lot for data management tasks.

Ⓐ Start ArcMap and open a blank map.

Inside the ParkSite folder, we'll create a new folder for the project. Eventually, it will contain a few geodatabases and some other items, but in this lesson, it will just have one geodatabase with the starting data for the project.

Ⓑ In the Catalog window, expand Folder Connections, and then expand the C:\UGIS folder.

Ⓒ Right-click the ParkSite folder. On the context menu, point to New and choose Folder.

Ⓓ Name the folder **AnalysisData** and press Enter.

Ⓔ Right-click the AnalysisData folder, point to New and choose File Geodatabase.

Ⓕ Rename the geodatabase **ReadyData** and press Enter.

Every map document has a default geodatabase, to which ArcMap directs geoprocessing output data (see page 97). Unless otherwise specified, the default geodatabase is located in your My Documents\ArcGIS folder.

Ⓖ Under AnalysisData, right-click the ReadyData geodatabase and choose Make Default Geodatabase.

In the Catalog window, the name of the geodatabase is boldfaced and a "home" is added to the icon (Figure 4-1). Right now, ReadyData is just an empty container. In the next exercise, we'll start to fill it with data.

Figure 4-1

ReadyData is now the default geodatabase.

Project database considerations

For our simple database, we considered the following issues:

Data format: We put all the data into file geodatabase format. The geodatabase is the standard Esri spatial data format, and it has many advantages, large and small, over the shapefile format. We decided not to use feature datasets ⬜. Feature datasets are only required to support advanced data structures (topologies and networks) that aren't discussed in this book. They can also be used as thematic organizers (they're used this way in the ESRI geodatabase in the source data for this project), but we won't have enough feature classes in our project database to make that worthwhile.

Inputs and outputs: An analysis project has input and output data. We'll store our input, or starting, data in one geodatabase and our output data in a separate geodatabase. This isn't the only way to do it. We could store both the input and output data in the same geodatabase. The reason we're doing it this way is because eventually we might want to run variations on the analysis. This would lead to multiple sets of results, where each set could consist of several feature classes. Keeping track of all that data in one geodatabase would be confusing.

Naming conventions: Conventions aren't right or wrong in and of themselves. Ideally, they should be easy to remember and easy to apply, but the only real sign that a convention is "wrong" is that you don't stick to it. Our conventions are:

- To separate word forms with capital letters, rather than spaces or underscores. For example, *VacantParcels* rather than *Vacant_Parcels* or *Vacant Parcels*. (This typographic convention is called "Pascal case" or "Camel case.")

- To use natural singulars and plurals. For example, *LARiver* (one river) and *MajorRoads* (many roads).

- To use complete names, not abbreviations. For example, *MajorRoads*, not *MjrRds*.

- To prefer descriptive names to names that reflect software procedures (the ArcMap default). For example, *ExcludedParkAreas* is better than *Parks_Buffer_Dissolve_Erase*.

Coordinate system: Feature classes in a geodatabase don't have to be in the same coordinate system. (Feature classes in the same feature dataset do have to be in the same system.) Since ArcMap projects data on the fly, you don't have to worry too much about mixing and matching datasets. One reason, however, to keep all the data in the same coordinate system is that we don't have to think about making sure the data frame is set to the appropriate system. Also, it's considered better practice to make spatial edits to data in its native coordinate system than when it's projected on the fly. We'll project all the data in our database to the State Plane California Zone 5 coordinate system.

Maintenance and documentation: We don't have to worry about updates, security, quality control, or other issues that apply to keeping a large database up and running. For us, maintaining the database means keeping it uncluttered. In the course of an analysis project, it's common to create datasets that are an important link in a chain of processes, but that have no particular value once the end of the chain is reached. If you don't remove this "intermediate" data from the geodatabase, it accumulates quickly, often making it hard to distinguish critical outputs and results from byproducts. (This is especially true for others who may want to look at or use your data, but it's true for you as well.) In addition to keeping the database trim, we also want to keep it reasonably well documented. Output datasets that we intend to keep should have their item descriptions updated to reflect the contents and purpose of the data.

Types of geodatabases

A geodatabase can be one of three types:

A **file geodatabase**, with extension .gdb, is stored as a collection of system files in a folder. It has no total size limit. Each dataset within the geodatabase has a default size limit of one terabyte, which can be increased to 256 terabytes. Multiple users can concurrently edit different datasets within the geodatabase.

A **personal geodatabase**, with extension .mdb, is a Microsoft® Access® database. It has a total size limit of two gigabytes. Only one user at a time can edit the database. File and personal geodatabases serve the same purposes; the personal geodatabase is just an older version of the technology.

An **ArcSDE geodatabase** is stored in a relational database system such as Oracle®, Microsoft SQL Server™, PostgresSQL, or IBM® DB2® and is managed through ArcSDE® software. (SDE stands for Spatial Database Engine.) Its size limits are determined by the database system in which it resides. ArcSDE geodatabases support concurrent editing on individual datasets and provide database administration tools.

Exercise 4b: Project a feature class

Our second goal is to populate the geodatabase with starting data for the analysis. As needed, we'll convert the data format from shapefile to geodatabase and project it to the State Plane coordinate system.

1)   **Open the data requirements table.** Having this table open may help you keep the various preparation tasks in mind. As you work through the lesson, you can refer to it or not as you like.

- ❹ Using Windows Explorer, open the data requirements table from the MapsAndMore folder.
  - ▷ If your table isn't up-to-date, download the results for Lesson 3 from the Understanding GIS web site.

Most of the datasets need some preparation. Actually, they all do, if we count format conversion and projection as preparation tasks. Let's start with a dataset that doesn't have anything listed in the PREPARATION column. The *Mjr_rd* feature class in Row 13 is one.

- ❸ Minimize the data requirements table.

2)   **Open the Project tool.** First of all, let's check the coordinate system of the *Mjr_rd* feature class to make sure it needs to be projected.

- ❹ In ArcMap, in the Catalog window, expand the SourceData folder, the ESRI geodatabase, and the Transport feature dataset.
- ❸ Right-click *Mjr_rd* and choose Properties.

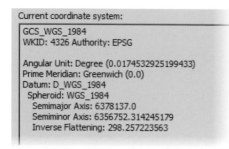

Current coordinate system:

GCS_WGS_1984
WKID: 4326 Authority: EPSG

Angular Unit: Degree (0.0174532925199433)
Prime Meridian: Greenwich (0.0)
Datum: D_WGS_1984
  Spheroid: WGS_1984
    Semimajor Axis: 6378137.0
    Semiminor Axis: 6356752.314245179
    Inverse Flattening: 298.257223563

Figure 4-2

Figure 4-3

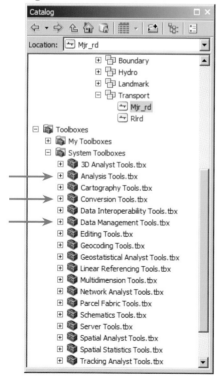

C In the Feature Class Properties dialog box, click the XY Coordinate System tab (Figure 4-2).

The coordinate system is GCS_WGS_1984. We'll project it to the State Plane system. This isn't an on-the-fly projection of a layer in a data frame, but rather a geoprocessing operation on the feature class itself that converts all the WGS_1984 angular coordinates to State Plane linear coordinates. We need a geoprocessing tool to do it.

D Close the Feature Class Properties dialog box.

E In the Catalog Window, at the same level as Folder Connections, but underneath it, expand the Toolboxes folder. In the Toolboxes folder, expand System Toolboxes.

The system toolboxes contain the geoprocessing tools that come with ArcGIS. There are a lot of them, but most are pretty specialized. The ones most useful to our project (and to much typical GIS work) are the analysis, conversion, and data management tools (Figure 4-3).

F Expand Data Management Tools. Expand the Projections and Transformations toolset, and then the Feature toolset.

As shown in Figure 4-4, toolboxes contain toolsets, which contain tools (and sometimes other toolsets).

Figure 4-4

G Right-click the Project tool (marked with a hammer icon) and choose Open.

H In the lower right corner of the Project tool dialog box, click Show Help (Figure 4-5).

If you click in a parameter, the Help panel will update with contextual help.

This button opens the ArcGIS Desktop Help System.

Figure 4-5

This button opens and closes the side Help panel.

Most tools look basically the same as this one: a dialog box with several settings, called parameters. Required parameters, such as the input and output datasets, are marked with a green dot.

The Help panel summarizes the tool's function. This tool takes an input dataset and creates a new output dataset in whatever coordinate system you choose.

I Optionally, click Hide Help to close the Help panel.

3) Project the *Mjr_rd* feature class. Now we'll fill out the parameters and run the tool. The first thing to specify is the input dataset, the *Mjr_rd* feature class.

A In the Catalog window, scroll up to ESRI.gdb in the SourceData folder. Under Transport, drag and drop *Mjr_rd* from the Catalog window to the Input Dataset or Feature Class parameter of the Project tool (Figure 4-6).

Figure 4-6

Once you drop the input dataset in place, ArcMap can fill out some of the other parameters for you. It reads the input coordinate system (GCS_WGS_1984) from the metadata and directs the output to the default geodatabase (Figure 4-7). This is where we want the data to go, but it's not the feature class name we want.

Figure 4-7

B In the Output Dataset or Feature Class box, highlight the entire path and press the Delete key. In its place, type **MajorRoads**.

C Press the Tab key.

The output path to the ReadyData geodatabase is restored.

D Next to the Output Coordinate System parameter, click the Spatial Reference Properties button.

E In the Spatial Reference Properties dialog box, make sure the XY Coordinate System tab is selected.

F Expand the Projected Coordinate Systems folder, the State Plane folder, and the NAD 1983 (US Feet) folder.

G Scroll down and click NAD 1983 StatePlane California V FIPS 0405 (US Feet) (Figure 4-8).

Figure 4-8

The bottom half of the dialog box shows the coordinate system details. The map projection is Lambert Conformal Conic. The values for the central meridian and standard parallels focus the coordinate system on a particular swath of southern California, pushing virtually all distortion outside the area of interest. Notice that the underlying geographic coordinate system is GCS_North_American_1983 (Figure 4-9).

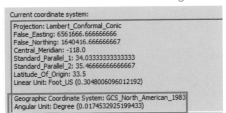

Figure 4-9

North American 1983 is the ArcGIS name for the geographic coordinate system based on the North American Datum of 1983 (NAD83).

H Click OK on the Spatial Reference Properties dialog box.

Because the input and output datasets have different geographic coordinate systems—WGS_1984 and NAD_1983 respectively—a geographic transformation has been listed automatically. When you project layers on the fly, you can bypass geographic transformations, but the Project tool makes you do them.

The default transformation chosen by ArcMap is usually the best choice, but there may be times when you want to use a different one; for example, to conform to standard practice in your organization. In this case, we'll switch to another suitable transformation. For more information about geographic transformations, see the topic *Datums* in Lesson 3.

Figure 4-10

I Click the Geographic Transformation drop-down arrow.

J In the drop-down list of available transformations (which includes only relevant options), choose NAD_1983_To_WGS_1984_5.

Remove this transformation

On the dialog box, both transformations are now listed.

K Click WGS_1984_(ITRF00)_To_NAD_1983 to select it and click the Delete ☒ button to remove it (Figure 4-10).

The name of the transformation we're using, NAD_1983_To_WGS_1984_5, looks as if it might be the wrong way around: we want the *Mjr_rd* feature class to be transformed from WGS1984 to NAD1983. But in fact the transformation works both ways, and the appropriate direction is set according to the input and output coordinate systems.

L Check your Project tool settings against Figure 4-11. Click OK.

When the process is done, a notification appears and fades in the lower right corner of your monitor, and a new layer is added to the map. A new feature class with the coordinate system you specified has also been saved to the ReadyData geodatabase.

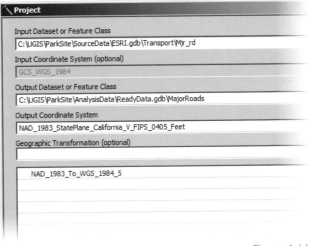

Figure 4-11

M In the Catalog window, expand the AnalysisData folder, if necessary, and expand ReadyData.gdb (Figure 4-12).

New projected feature class

The original *Mjr_rd* feature class remains unchanged in the ESRI geodatabase.

Figure 4-12

N Confirm that the coordinate system of *MajorRoads* is State Plane California Zone V, and that its geographic coordinate system is GCS_North_American_1983.

▷ How? Use the feature class properties (XY Coordinate System tab) or the layer properties (Source tab).

O From the ArcMap main menu, choose Geoprocessing → Results.

P In the Results window, expand Current Session and Project. Under Project, expand Inputs and expand Messages (Figure 4-13).

The Results window keeps a log of which geoprocessing operations have been run, at what times, and with what parameters. Results are especially useful when a process fails because the error is reported in the Messages section.

Double-clicking the tool icon (in this case, the hammer) next to the tool name will reopen the tool with the parameters you used. This can be a convenient way to rerun the tool if you want to change a parameter.

By default, geoprocessing results are stored for two weeks. You can change this value by choosing Geoprocessing → Geoprocessing Options from the main menu.

Q Close the Results window.

Figure 4-13

In the next exercise, you'll add two more feature classes to the ReadyData geodatabase.

Exercise 4c: Convert feature classes

In this exercise, we'll project the *block_centroids* and *Parkland* feature classes. We'll learn a couple of useful things. The first is how to set the geoprocessing environment. Environment settings are default values or conditions applied automatically to all geoprocessing operations. The second thing is how to use a batch tool that can process multiple inputs.

1) **Set the geoprocessing environment.** There are many environment settings, but we'll concentrate on the output coordinate system.

Figure 4-14

Click here (or on the double arrows) to expand the heading

Ⓐ From the main menu, choose Geoprocessing ➔ Environments.

Ⓑ In the Environment Settings box, click Output Coordinates to expand the heading (Figure 4-14).

Ⓒ For the Output Coordinate System parameter, click the drop-down arrow and choose "Same as Layer 'Major Roads.'"

Since *MajorRoads* is in State Plane California Zone V, this becomes the coordinate system environment setting.

Ⓓ Click the Geographic Transformations drop-down arrow.

Ⓔ Scroll down about three-quarters of the way and select NAD_1983_To_WGS_1984_5 (Figure 4-15).

The transformation name is added to a box, as shown in Figure 4-16.

Ⓕ Click OK.

By making our desired output coordinate system an environment setting, we tell ArcMap to make sure that whenever a geoprocessing tool runs, the output is projected to that system.

Figure 4-15

Figure 4-16

Figure 4-17

2) Convert multiple feature classes. The *Parkland* feature class is in the WGS 1984 geographic coordinate system and needs to be projected. The *block_centroids* feature class is a shapefile in the WGS 1984 system. It needs to be both converted to geodatabase format and projected.

A Scroll down in the Catalog window. Under System Toolboxes, expand Conversion Tools, and then expand To Geodatabase.

B Right-click the Feature Class to Geodatabase (multiple) tool (Figure 4-17) and choose Open.

▷ Or open the tool by double-clicking it.

C In the Catalog window, scroll up to the SourceData folder.

D Under the ESRI geodatabase, expand the Landmark feature dataset.

E Drag and drop *Parkland* from the Catalog window to the Input Features parameter of the tool.

The feature class is added as the first item in a list underneath the Input Features parameter.

F In the Catalog window, expand the census folder.

G Drag and drop *block_centroids.shp* to the tool's input parameter.

H In the Catalog window, in the AnalysisData folder, click ReadyData.gdb to highlight it.

I Drag and drop ReadyData into the tool's Output Geodatabase parameter.

J Confirm that your tool dialog box looks like Figure 4-18 and click OK.

The tool runs and informs you when it's done.

3) Inspect the results. Let's see what we've accomplished.

A In the Catalog window, expand the ReadyData geodatabase, if necessary.

It should contain three feature classes: *block_centroids*, *MajorRoads*, and *Parkland*.

B Check the coordinate systems of the *block_centroids* and *Parkland* feature classes.

Figure 4-18

This tool copied both input feature classes to the ReadyData geodatabase, converting the *block_centroids* shapefile to geodatabase format in the process. Thanks to the environment setting, both feature classes were also projected to the State Plane coordinate system. As long as that setting remains in place—which it will, within the current map document, until you change it—any geoprocessing tool you run will also cause the input dataset to be projected. As usual with geoprocessing operations, the outputs are new feature classes and the original inputs are unchanged.

You probably noticed that the tool icon 📃 looked like a scroll of paper. Tools with this icon are script tools and don't always conform to the same protocols as built-in tools 🔧. For example, the tool's output geodatabase parameter didn't default to ReadyData. Nor were the converted feature classes added as layers to the map.

4) Rename feature classes. We'll rename the new feature classes according to our convention.

- **Ⓐ** In the Catalog window, under ReadyData, right-click *block_centroids* and choose Rename.
  - ▷ Or, click the feature class name to highlight it and click again to edit it.

The name is surrounded by a black box, making it editable.

- **Ⓑ** Rename it **BlockCentroids**.
- **Ⓒ** Rename *Parkland* to **Parks** (Figure 4-19).

This is a good time to save the map document.

- **Ⓓ** On the Standard toolbar, click the Save button 💾.
- **Ⓔ** Navigate, if necessary, to the MapsAndMore folder. Name the map document **Lesson4** and click Save.

In the Catalog window, the home folder changes to ParkSite\MapsAndMore.

- **Ⓕ** If you're continuing to the next exercise now, leave ArcMap open; otherwise, exit ArcMap and close the data requirements table.

Figure 4-19

Tool types

The tools at your disposal depend on your ArcGIS product level (Basic, Standard, or Advanced) and on whether you have extension products installed. Tools have different icons according to their type. Built-in tools 🔧 are written in a compiled programming language and can't be modified. Script tools 📃 are mostly written in the Python® scripting language. You can view, copy, and edit the source code. Many of the script tools that come with ArcGIS adapt built-in tools to work with multiple inputs. Model tools 🔧 encapsulate sequences of data processing operations. In Lesson 7, you'll create a model tool that packages the entire analysis project as a tool.

Exercise 4d: Prepare the city and county data

Now we'll move on to data that needs more preparation. In the requirements table, we listed *City_ply* and *County* as data needed for the final map. We made notes to reduce their extent because both feature classes cover the entire United States.

Processing data to get rid of extra features or attributes is a matter of choice. We could just as well leave the datasets intact and instead apply appropriate definition queries to the *City_ply* and *County* layers in the final map. One point in favor of small datasets is that they generally display and process faster. Likewise, an attribute table is easier to read if it's not cluttered with unnecessary attributes. If we later decide to share the project data with others, it will also be convenient to have a smaller package.

We need to be careful in deciding which features and attributes are "unnecessary," because once we get rid of them, they're gone—gone, at least, from that dataset. We'll still have the original source data as a backup, so we can recover from mistakes. We probably wouldn't delete features or attributes that weren't preserved in another feature class.

1) **Add data to the map.** We'll add the *City_ply* and *County* feature classes as layers to the map. Then we'll select the features we want to keep and copy the selections to new feature classes.

- Ⓐ If necessary, start ArcMap and open Lesson4 from the list of recent maps.
- Ⓑ In the Catalog window, in the SourceData folder, under ESRI.gdb, expand the Boundary feature dataset.
- Ⓒ Click on the Boundary feature dataset name to highlight it.
- Ⓓ At the top of the Catalog window, click the Toggle Contents Panel button 🔡.

The Catalog window splits into two panes. You can use the bottom pane to drag multiple datasets onto the map.

- Ⓔ In the bottom pane, hold down the Control key and click the *City_ply* and *County* feature classes to highlight both (Figure 4-20).
- Ⓕ Drag and drop the datasets into the map window.
  - ▷ If you get a geographic coordinate systems warning, check the box that says "Don't warn me again in this session" and click Close.

The two layers are added to the map.

- Ⓖ In the Table of Contents, click in some empty white space to unselect the layers.

Figure 4-20

ⓗ If necessary, drag *City_ply* above *County*.

ⓘ Optionally, change the symbols to colors that suit you.

ⓙ Remove the *MajorRoads* layer from the map.

▷ How? In the Table of Contents, right-click the layer name and choose Remove.

2) Select counties. In this step, we'll select the county features that we want to keep.

ⓐ Turn off the *City_ply* layer.

ⓑ In the Table of Contents, right-click *County* and choose Label Features.

The counties are labeled with their names. Label positions change dynamically with the map scale and extent, so yours may be placed differently.

ⓒ If necessary, pan the map to put Los Angeles County roughly in the middle of the display (Figure 4-21).

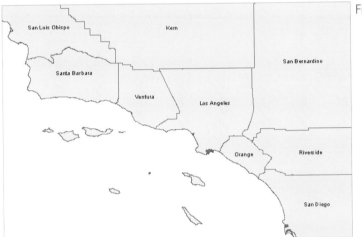

Figure 4-21

ⓓ On the Tools toolbar, click the Select Features tool 🢬. Click anywhere inside Los Angeles county to select it.

The selected feature is outlined in bright blue (Figure 4-22). This is probably the only county feature we'll need for the final map, but to be safe, we'll select its neighbors as well. We'll do this with a spatial query.

ⓔ From the main menu, choose Selection → Select By Location.

Spatial queries evaluate spatial relationships among features. These relationships include proximity, containment, adjacency, and intersection. Typically, features in one layer (the target layer) are selected according to their relationship to features in another layer (the source layer). Sometimes, as here, the target and source layers are the same.

Figure 4-22

Features in this layer will be selected according to their spatial relationship to features in this layer

Figure 4-23

F In the Select By Location dialog box, under Target Layer(s), check the box next to *County*.

G Click the Source Layer drop-down arrow and choose *County*.

H At the top of the dialog box, click the Selection method drop-down arrow and choose "add to the currently selected features in."

We want to keep our currently selected feature, Los Angeles County, and select more features.

I Confirm that the Spatial selection method for target layer feature(s) drop-down list is set to "intersect the source layer feature."

If feature boundaries touch each other, ArcMap considers this to be intersection, so this selection method will select counties that border Los Angeles. (Alternatively, you could choose the method "touch the boundary of the source layer feature" to get the same result.)

J Compare your dialog box to Figure 4-23 and click OK.

On the map, Los Angeles, Ventura, Kern, San Bernardino, and Orange counties are selected.

3) Turn off fields. The *County* layer will serve a purely cartographic purpose. That means the only attributes we need are those we might want for labeling or symbology.

A In the Table of Contents, right-click the *County* layer and choose Open Attribute Table.

B Scroll quickly across the attributes.

There are about fifty demographic and economic attributes— none of which we need. Removing them will tailor our data to its cartographic purpose.

C Close the table.

D Open the layer properties for the *County* layer and click the Fields tab (Figure 4-24).

Figure 4-24

The scrolling box on the left lists the fields in the table. Each field's check box can be unchecked to hide the field. Field visibility is a layer property that you can turn on and off as you like—a hidden field isn't deleted. Nevertheless, the setting does affect the way data is copied. When you copy the feature class, hidden fields will be left out.

E Above the list of fields, click the Turn all fields off button ▦.

F Click the boxes next to NAME and STATE_NAME to turn just these two fields back on (Figure 4-25). Click OK.

G Open the *County* attribute table to see the effect, then close the table.

H Save the map.

Figure 4-25

4) Copy the selected features to a new feature class. A geoprocessing tool called Copy Features will copy the five selected features and their attributes to a new feature class. Instead of hunting for the tool among the System toolboxes, we'll make ArcMap get it for us.

A On the right side of the ArcMap application window, click the Search tab to open the Search window.

▷ If the tab isn't showing, click the Search button 🔍 on the Standard toolbar.

B Near the top of the window, click Tools (Figure 4-26).

Figure 4-27

Figure 4-26

The system toolboxes are displayed in the window.

C In the search box, type **copy**.

As you type, ArcMap presents a list of possible matches in a drop-down box.

D In the drop-down box, click "copy features (data management)" (Figure 4-27).

In the Search window, Copy Features (Data Management) is returned at the top of a list of tools.

E In the list, click on the tool's name to open the tool.

G In the Copy Features dialog box, click the Input features drop-down arrow and choose *County*.

You can choose input features from a drop-down list when there are layers in the map document.

G In the Output Feature Class parameter, highlight and delete the entire path. Type **Counties** and press Tab (Figure 4-28).

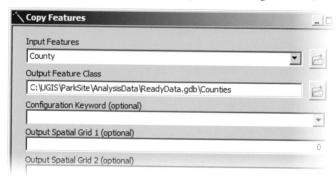

Figure 4-28

H Click OK to run the tool.

When the tool is done, a *Counties* feature class is added to ReadyData and a corresponding layer is added to the map.

I In the Catalog window, under ReadyData, right-click *Counties* and choose Properties.

J In the Feature Class Properties dialog box, click the Fields tab (Figure 4-29).

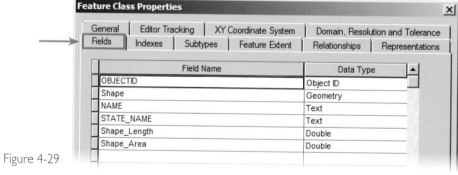

Figure 4-29

The only fields copied from the *County* layer are NAME and STATE_NAME. The other four are created automatically by ArcMap.

K Click the XY Coordinate System tab.

Again, your environment setting has ensured that the data was not only copied but also projected to state plane coordinates.

L Close the Feature Class Properties dialog box.

Ⓜ In the Table of Contents, remove the (old) *County* layer.

Ⓝ Right-click the (new) *Counties* layer and choose Zoom To Layer.

The *Counties* layer consists of the five counties you selected. The islands are part of the Los Angeles and Ventura county features. Features with discontinuous geometry (islands are a common example) are called "multipart" features. If you click on an island with the Identify tool, you'll see the entire county flash green.

5) Select cities. As with the counties, we don't know exactly which cities we need for the final map. We'll make an interactive selection that takes in an indefinite, but ample, number of features.

Ⓐ In the Table of Contents, turn on the *City_ply* layer.

Ⓑ Drag *City_ply* above *Counties*.

Ⓒ Right-click *City_ply* and choose Selection → Make This The Only Selectable Layer.

Ⓓ On the Tools toolbar, click the Select Features tool 🔍.

Ⓔ Use the tool to drag a selection box similar to the one in Figure 4-30. (Your box can be bigger or smaller.)

When you release the mouse button, cities entirely or partially within the box are selected (Figure 4-31).

Figure 4-30

Figure 4-31

Ⓕ At the top of the Table of Contents, click the List By Selection button 🔳.

In the example, 262 cities have been selected (Figure 4-32). It doesn't matter if you selected more or fewer.

Ⓖ At the top of the Table of Contents, click the List By Drawing Order button 🔖.

Figure 4-32

Figure 4-33

6) **Turn off fields.** As we did with the counties, we'll get rid of attributes we don't need.

Ⓐ Open the layer properties for the *City_ply* layer and click the Fields tab, if necessary.

Ⓑ Turn off all the fields. ▣

Ⓒ Click the check boxes next to NAME and ST (an abbreviation for State) to turn these two fields back on (Figure 4-33).

Ⓓ Click OK on the Layer Properties dialog box.

7) **Copy features.** Now we'll copy the selected cities to a new feature class in ReadyData.

Ⓐ Click the Search tab to open the Search window.

 ▷ Hovering over the tab will also open the window as long as the ArcMap application window is active.

The Copy Features tool is still at the top of the list.

Ⓑ Click the tool's name to open it.

Ⓒ In the Copy Features dialog box, click the Input Features drop-down arrow and choose *City_ply*.

Ⓓ In the Output Feature Class parameter, clear the path. Type **Cities** and press Tab.

Ⓔ Compare your tool dialog box to Figure 4-34 and click OK.

Figure 4-34

When the tool is finished, a *Cities* feature class is created in ReadyData and a layer is added to the map.

Ⓕ In the Catalog window, confirm that you have a *Cities* feature class in ReadyData.

Ⓖ In the Table of Contents, remove the (old) *City_ply* layer.

Ⓗ Open the attribute table for the (new) *Cities* layer.

The table contains the two attributes you kept, plus the software-managed attributes (Figure 4-35).

Figure 4-35

Table

Cities

	OBJECTID *	Shape *	NAME	ST	Shape_Length
▶	1	Polygon	Santa Paula	CA	94746.082904
	2	Polygon	San Buenaventura (Ventura)	CA	404046.495289
	3	Polygon	Ojai	CA	75894.965569
	4	Polygon	Moorpark	CA	116096.162704
	5	Polygon	Fillmore	CA	72935.368563
	6	Polygon	Santa Clarita	CA	417446.017356
	7	Polygon	San Fernando	CA	36618.714967
	8	Polygon	Frazier Park	CA	73665.456345

❶ Close the table.

❷ Save the map document 💾.

❸ If you're continuing to the next exercise now, leave ArcMap open; otherwise, exit ArcMap.

Using the Search window

The Search window lets you search for map documents and data as well as tools. Tools are indexed automatically by the software, but to search for maps and data, you have to index the workspaces that you want searched. To create an index, click the Search Options button ⚏ at the top of the Search window. In any search, you can type keywords into the search box to help you find items that you don't know by name. Keyword searches use information in the item's metadata.

Exercise 4e: Prepare the river data

In Row 5 of the data requirements table, we made a note to dissolve the *LARiver* dataset. This will combine the river's 265 features into one, simplifying the task of creating a buffer around the river in Lesson 6. The Dissolve tool creates a new feature class. When we run the tool, we'll convert the *LARiver* shapefile to a geodatabase feature class as part of the operation.

1) **Dissolve the river.** A few common geoprocessing tools, including Dissolve, can be accessed from the main menu.

Ⓐ If necessary, start ArcMap and open Lesson4 from the list of recent maps.

Ⓑ From the main menu, choose Geoprocessing ➔ Dissolve.

Ⓒ In the Catalog window, under SourceData, expand the City of LA folder.

Ⓓ Drag *LARiver.shp* to the tool's Input Features parameter.

The output feature class location defaults to ReadyData.

Dissolving features

You can dissolve a feature class either by geometry or by an attribute. If you dissolve by geometry, common boundaries between features disintegrate. The number of output features depends on whether the "Create multipart features" box is checked. If it's checked (the default), you'll get one output feature with discontinuous geometry. If you uncheck it, you'll get a unique, singlepart feature for each spatially distinct area. In either case, the output table has no attributes except for those maintained by ArcMap.

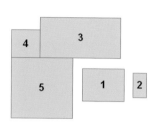

Five input features will be
dissolved by geometry

☑ Create multipart features (optional)

The output is one
multipart feature

☐ Create multipart features (optional)

The output is three
singlepart features

If you dissolve by an attribute, common boundaries between features disintegrate only if the features have the same attribute value. Again, the number of output features depends on whether multipart features are created. If so, all input features with the same attribute value will become one output feature (with discontinuous geometry). If not, you'll get a unique feature for each spatially distinct area. The attribute being dissolved on is preserved in the output table.

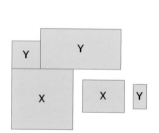

Five input features will be
dissolved on a TYPE attribute

☑ Create multipart features (optional)

The output is two
multipart features: one X
feature and one Y feature

☐ Create multipart features (optional)

The output is four
singlepart features: two X
features and two Y features

In addition to the attribute being dissolved on, you can set a "statistics field" to summarize the values of dissolved features for another attribute. In this example, the features are dissolved by TYPE and SUM_ACRES is a statistics field (based on an ACRES field in the original table) that calculates the total acreage of the dissolved features.

E Change the output feature class name from LARiver_Dissolve to **LARiver**.

▷ You can use the method of clearing the path and typing the new name, or just backspace over the part of the name you want to remove.

F In the Dissolve Fields area, check the NAME box.

Checking this box preserves the NAME field, which we want to keep, in the output attribute table. It also affects the way the tool works. Instead of creating a single output feature, ArcMap will create a feature for each unique value in the checked field. In our case, it comes to the same thing because the NAME field only has one value ("Los Angeles River"). If we checked a different field, however, the output feature class would have more than one feature.

G Compare your dialog box to Figure 4-36 and click OK.

When the tool is done, a feature class is added to ReadyData and a layer is added to the map.

H Open the attribute table for the *LARiver* layer (Figure 4-37).

Figure 4-36

It contains one record. Except for the NAME field you checked, the attributes from the original table have been dropped.

I Close the attribute table.

J Optionally, symbolize the river in a shade of blue and increase its width.

Exercise 4f: Prepare the park data

The *Parks* feature class is already in the ReadyData geodatabase, but we still have the preparation task of loading data into it. That's because this dataset is missing the two park features that we saw in the *NewParks* shapefile.

Figure 4-37

1) **Add data to the map.** Let's remind ourselves which two parks we're talking about.

A From the ReadyData geodatabase, add *Parks* to the map. Optionally, symbolize the layer in a shade of green.

B Under SourceData, expand the ParkData folder and add *NewParks.shp*. Optionally, symbolize this layer in a different shade of green.

© Zoom to the *NewParks* layer (Figure 4-38).

2) Compare attributes. We want to get these two *NewParks* features, along with their attributes, into the *Parks* feature class. We face a complication in that the datasets have different attributes.

Ⓐ Open the *Parks* attribute table.

It has 1,209 records. Its user-managed attributes are:

NAME   TYPE   ACRES

The TYPE attribute describes whether the park is a local, state, or national park.

Ⓑ Open the *NewParks* attribute table.

Its user-managed attributes are:

NAME   CATEGORY   ADDRESS   TELEPHONE   HOURS

Both tables have a NAME field. CATEGORY is similar to the TYPE field in the *Parks* table in that it identifies the parks as state parks. The other fields don't have equivalents in the *Parks* table.

© Close the table window.

3) Load data. We'll load the data and see how ArcMap handles the attribute mismatches.

Ⓐ In the Catalog window, under ReadyData, right-click *Parks* and choose Load → Load Data.

The Simple Data Loader wizard opens.

Ⓑ On the wizard panel, click Next to advance.

Figure 4-38

Figure 4-39

© Next to Input data, click the Open Folder button.

Ⓓ In the Open GeoDatabase dialog box, browse to the SourceData\ParkData folder. Click *NewParks.shp* and click Open.

Ⓔ On the wizard panel, click the Add button.

NewParks.shp is the data that we want to load (Figure 4-39).

Ⓕ Click Next to advance.

This panel confirms that the data will be loaded into an existing geodatabase (ReadyData.gdb) and into the feature class *Parks*. Neither setting can be changed.

Ⓖ Click Next to advance.

In the next panel, you specify how to handle attributes. ArcMap assumes that you want NAME values from the source feature class (*NewParks*) to be copied into the NAME field of the target feature class (*Parks*). It doesn't make any other assumptions because the other field names don't match (Figure 4-40).

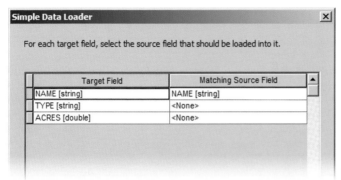

Figure 4-40

❽ In the second row (where TYPE is the Target Field value), click <None> in the Matching Source Field column.

A drop-down list appears with the attributes from *NewParks*.

❾ In the list of attributes, click CATEGORY (Figure 4-41).

Target and Source field names don't have to match, but they must be of the same data type (string, for example). For more information, see the topic *Field data types*.

Figure 4-41

Values from the CATEGORY field in *NewParks* will be copied into the TYPE field of *Parks*.

There's no ACRES field (by this or any other name) in the *NewParks* table, so leave the Matching Source Field in the third row set to <None>. No values will be copied into the ACRES field in *Parks*.

The ADDRESS, TELEPHONE, and HOURS information in *NewParks* won't be loaded, because *Parks* has no appropriate fields to accept the values.

Field data types

One of the essential properties of a field in an attribute table is its data type, which defines the type of information the field can hold. When a new field is added to a table, its data type is specified and can't be changed thereafter. There are several field data types, but those used most often are text (sometimes called string) and the four numeric types: short integer, long integer, float, and double.

Text fields are for descriptions, codes, noncomputational numbers (such as postal codes or telephone numbers), and the like. The default length of a text field is 50 characters. If you know that your entries will be shorter—for example, if you're storing two-letter state abbreviations—you should set the length property accordingly. In the interest of saving space, you can also use codes in place of long descriptions. In a geodatabase, codes and their associated values can be managed through a so-called attribute domain (not discussed in this book).

For numeric data, use short or long integers when the values are whole numbers, such as population or number of items sold. Whether to use the short or long integer type depends on how large the numbers may be (see the table below). Use floats or doubles when the values are fractional. Floats store six digits with precision (for example, 12345.6 or 1.23456). Numbers with more digits are stored with some rounding, which may affect calculations slightly. Doubles are stored with fifteen digits of precision. Floats are suitable for most statistical calculations, such as population density or average income. Doubles are recommended for fields storing geographic measurements and calculations. ArcMap uses the Double type in its Shape_Length and Shape_Area fields.

The following table summarizes the field data types available for geodatabase feature classes. Not all are available for shapefiles.

Field Type	Stores
Text	Letters, numbers, special characters
Short integer	Whole numbers from -32,768 to 32,767 (uses 2 bytes)
Long integer	Whole numbers from -2,147,483,648 to 2,147,483,647 (uses 4 bytes)
Float	Fractional numbers: precise to 6 digits, then rounded (uses 4 bytes)
Double	Fractional numbers: precise to 15 digits, then rounded (uses 8 bytes)
Date	Dates and/or times
BLOB	Binary large objects, including special feature types like geodatabase annotation
Raster	Small images
GUID	Unique feature/record identifier for features in distributed geodatabases
ObjectID	Unique feature/record identifier
Shape	Feature geometry type (for example, point, line, or polygon)

J Confirm that the wizard panel looks like Figure 4-42.

▷ If you made a mistake, click Reset on the wizard panel and start again.

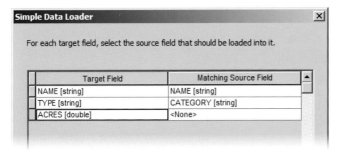

Figure 4-42

K Click Next to advance.

The default setting on this panel is to load all the source data. That's what we want.

L Click Next to advance.

M Click Finish on the summary panel.

When you load data, you don't get any messages. The operation happens quickly and invisibly: it's already done.

4) View results. Let's see what we accomplished.

A In the Table of Contents, turn off the *NewParks* layer.

The new features now show up in the *Parks* layer (Figure 4-43).

B Open the Parks attribute table and scroll to the bottom.

The table now has 1,211 records, instead of the 1,209 it had before. The last two records are the ones that were loaded. They have values in the NAME and TYPE fields, but not in the ACRES field (Figure 4-44).

C Close the table.

D Remove the *NewParks* layer from the Table of Contents.

Figure 4-43

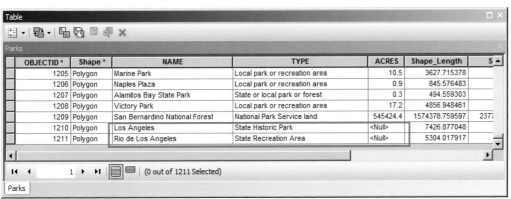

OBJECTID*	Shape *	NAME	TYPE	ACRES	Shape_Length	S
1205	Polygon	Marine Park	Local park or recreation area	10.5	3627.715378	
1206	Polygon	Naples Plaza	Local park or recreation area	0.9	845.576483	
1207	Polygon	Alamitos Bay State Park	State or local park or forest	0.3	494.559303	
1208	Polygon	Victory Park	Local park or recreation area	17.2	4856.948461	
1209	Polygon	San Bernardino National Forest	National Park Service land	545424.4	1574378.759597	2377
1210	Polygon	Los Angeles	State Historic Park	<Null>	7426.877048	
1211	Polygon	Rio de Los Angeles	State Recreation Area	<Null>	5304.017917	

Figure 4-44

Figure 4-45

5) **Make a spatial selection on the *Parks* layer.** It's not one of our preparation tasks to reduce the *Parks* data, but the spatial extent of this feature class is bigger than it needs to be. Getting rid of extra features may help speed up some analysis operations in Lessons 6 and 7.

A Zoom to the *Parks* layer.

B Turn off the *Cities* layer.

Those big national forests to the north are definitely outside our area of interest.

C From the ArcMap main menu, choose Selection → Select By Location.

D In the Select By Location dialog box, set the selection method, if necessary, to "select features from."

E Under Target Layer(s), check *Parks*.

F If necessary, click the Source Layer drop-down arrow and choose *LARiver*.

G Click the Spatial selection method for target layer feature(s) drop-down arrow and choose "are within a distance of the source layer feature" (the third entry in the list).

H Highlight and delete whatever numeric value is in the box underneath "Apply a search distance." Type **3**.

I Click the drop-down arrow in the adjacent box and set the units to Miles.

The spatial query will select parks within three miles of the river. If any bit of a park is within this distance, the feature will be selected. All the unselected features will be well out of our area of interest.

J Compare your dialog box to Figure 4-45 and click OK.

On the map, park features within three miles of the river are selected (Figure 4-46).

K At the top of the Table of Contents, click the List By Selection button 🔽 .

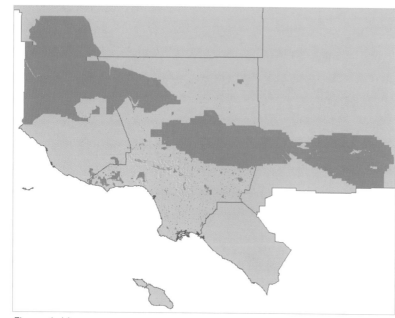

Figure 4-46

There should be 169 selected parks (Figure 4-47).

O Click the List By Drawing Order button %.

6) **Copy features.** Now we'll copy the selected parks, but we have to create a new output feature class. The *Parks* layer, with its 169 selected features, already references the *Parks* feature class in ReadyData, and ArcMap won't let it overwrite itself.

A In the Search window, locate and open the Copy Features tool.

B In the tool dialog box, set the Input Features to the *Parks* layer.

C Clear the Output Feature Class path, type **Parks2**, and press Tab.

D Compare your Copy Features dialog box to Figure 4-48 and click OK.

Figure 4-47

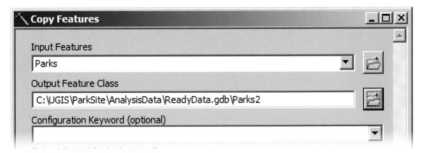

Figure 4-48

When the tool is finished, *Parks2* is created in ReadyData and a layer is added to the map.

E In the Table of Contents, turn off the (old) *Parks* layer.

F Resymbolize *Parks2* in any shade of green.

G Zoom to the *Parks2* layer and open its attribute table.

There should be 169 records.

H Close the table.

We don't need the *Parks* feature class any more, so we'll delete it from the geodatabase.

I In the Catalog window, under ReadyData, right-click *Parks* and choose Delete.

Make sure *Parks*—not *Parks2*—is highlighted in the Catalog window.

J Click Yes on the prompt to confirm your deletion.

The feature class is deleted from ReadyData and the corresponding layer is removed from the map. (You still have the original *Parkland* data in your SourceData folder, so you can recover from mistakes if you have to.)

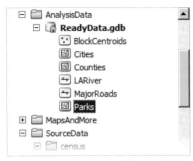

Figure 4-49

K In the Catalog window, right-click *Parks2* and choose Rename. Rename the feature class to *Parks* and press Enter (Figure 4-49).

The map layer pointing to this feature class is still called *Parks2*, which is fine.

L Save the map document.

M If you're continuing to the next exercise now, leave ArcMap open; otherwise, exit ArcMap.

Overwriting geoprocessing outputs

By default, ArcMap doesn't allow geoprocessing outputs to overwrite existing datasets. This is a useful safeguard, but if you need to change it, choose Geoprocessing Options from the Geoprocessing menu and check the box to "Overwrite the outputs of geoprocessing operations." (Changing this setting still won't allow you to overwrite a feature class with itself.)

Exercise 4g: Prepare the block group data

In Rows 8 and 9 of the data requirements table we identified some block group attributes—population density and age under 18—that we need to calculate from our existing data. We'll add new fields to the table, then calculate values into them.

1)   Add data to the map. First, we'll add the *block_groups* shapefile to the map and have another look at it before we copy it to ReadyData.

A If necessary, start ArcMap and open Lesson4.

B In the Catalog window, under SourceData, expand the census folder. Drag and drop *block_groups.shp* onto the map.

▷ If you get a Geographic Coordinate Systems Warning, close it.

C Zoom to the *block_groups* layer (Figure 4-50).

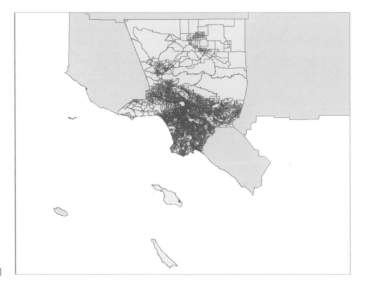

Figure 4-50

The spatial extent of the layer is Los Angeles County.

ⓓ Open the *block_groups* attribute table and scroll across the attributes.

We only need a handful of the two dozen or so attributes in the table. As with the parks, we didn't make a note to reduce this dataset, but it will be easier to work with the important attributes if we don't have to keep scrolling past the unimportant ones.

ⓔ Close the table.

2) **Turn off fields.** We'll turn off the fields we don't need.

ⓐ Open the layer properties for the *block_groups* layer and click the Fields tab, if necessary.

ⓑ Click the Turn all fields off button ⊡.

ⓒ Check the boxes for the following fields to turn them back on:
- FIPS (a unique user-managed identifier)
- TOTPOP_CY (total population)
- POP18UP_CY (population over 18)
- MEDHINC_CY (median household income)

ⓓ Confirm that your list of visible fields matches Figure 4-51. (You'll need to scroll down in your list.)

ⓔ Click OK.

3) **Import block groups.** Earlier, we used the Feature Class to Geodatabase (multiple) tool to get *BlockCentroids* and *Parks* into ReadyData. This time we'll use Feature Class to Feature Class, a similar tool that converts one feature class at a time. It has some advantages over Feature Class to Geodatabase (multiple), as well as over Copy Features, in that it lets you modify the properties of fields in the output table. The Feature Class to Feature Class tool is found in the Conversion Tools toolbox, but it can also be opened from a geodatabase's context menu.

ⓐ In the Catalog window, right-click the ReadyData geodatabase and choose Import → Feature Class (single).

The Feature Class to Feature Class tool opens.

Choose which fields will be visible
- ☐ FID
- ☐ Shape
- ☐ STCOFIPS
- ☐ TRACT
- ☐ BLKGRP
- ☑ FIPS
- ☐ POP2000
- ☐ POP2010
- ☑ TOTPOP_CY
- ☐ HHPOP_CY
- ☐ FAMPOP_CY
- ☐ GQPOP_CY
- ☐ TOTHH_CY
- ☐ AVGHHSZ_CY
- ☐ FAMHH_CY
- ☐ AVGFMSZ_CY
- ☐ TOTHU_CY
- ☐ OWNER_CY
- ☐ RENTER_CY
- ☐ VACANT_CY
- ☑ POP18UP_CY
- ☐ MEDAGE_CY
- ☑ MEDHINC_CY
- ☐ AVGHINC_CY
- ☐ AGGHINC_CY
- ☐ PCI_CY

Figure 4-51

B In the Input Features parameter, click the drop-down arrow and choose *block_groups*.

The output location is automatically set to ReadyData.

C In the Output Feature Class parameter, type **BlockGroups**.

D In the Field Map area, right-click the TOTPOP_CY field and choose Properties.

E In the Output Field Properties dialog box, change the name to **TOTPOP** and change the alias to **TOTAL POP** (with a space).

For more information about aliases, see the callout box *Field names and aliases*.

F Click the Type drop-down arrow and change the type from Long to Short (Figure 4-52). Click OK.

Figure 4-52

Every field has a data type that specifies what kind of information it holds. The Long and Short types both store integers. The largest value that can be stored by the Short type is 32,767. Since no block groups have a population that big, this type is adequate for our needs. See the topic *Field data types* for more information.

G Change the output field properties for the remaining fields as follows:

- For POP18UP_CY, change the name to **POP18UP** and the alias to **POP OVER 18**. Change the field type from Long to Short.
- For MEDHINC_CY, change the name to **MEDHINC** and the alias to **MEDIAN HH INCOME**. (Don't change the field type.)

H Compare your Feature Class to Feature Class dialog box to Figure 4-53 and click OK to run the tool.

When the tool is finished, a new feature class is created in ReadyData and a layer is added to the map.

I In the Table of Contents, remove the (old) *block_groups* layer.

J Open the attribute table for the (new) *BlockGroups* layer.

The table contains the attributes you kept and displays their aliases (Figure 4-54).

Figure 4-53

Figure 4-54

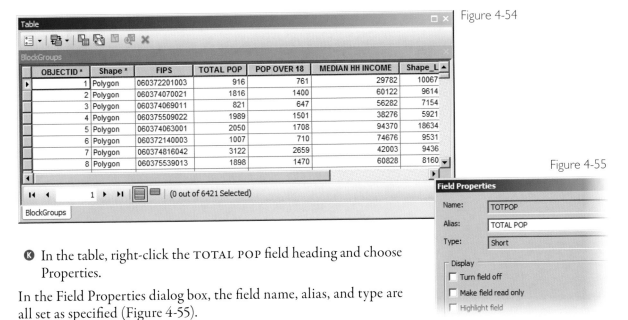

K In the table, right-click the TOTAL POP field heading and choose Properties.

In the Field Properties dialog box, the field name, alias, and type are all set as specified (Figure 4-55).

Figure 4-55

L Close the Field Properties dialog box, but leave the table open.

Field names in a geodatabase table are limited to 64 characters. They must begin with a letter and cannot include spaces. A few reserved words (for example, "Add," "Drop," and "Select") can't be used as field names. If you try to name a field with a disallowed name, you'll be prompted to rename the field. Field names in a shapefile table are limited to 10 characters. They have the same naming restrictions as geodatabase tables, except that there are no reserved words. Once set, the name of a field can't be changed.

An alias is a flexible and descriptive alternative name that can be displayed instead of a field's actual name. For fields in geodatabase feature classes, an alias can be set either as a feature class property or as a layer property. If set as a feature class property, it will be used whenever the feature class is added as a layer. If set as a layer property, it will be used only for the particular layer in which it is set. For fields in shapefiles, aliases are supported only as layer properties.

4) Calculate population density. Population density is the ratio of population to area. The SHAPE_AREA field stores the area of each block group in square feet. (This field was added automatically by ArcMap when we converted the block groups to geodatabase format.) We know the units are square feet because those are the measurement units specified by our coordinate system. Since we need to express density in terms of people per square mile, we're going to add two fields to the *BlockGroups* attribute table: one to store area in square miles and another to store population density expressed in square miles. Then we'll make the calculations for both fields.

Ⓐ At the top of the table, click the Table Options button ▦ . On the context menu, choose Add Field.

Ⓑ In the Add Field dialog box, for the name, type **SQMILES**.

Ⓒ For Type, click the drop-down arrow and choose Double.

Ⓓ In the Field Properties area, click in the white box next to Alias and type **SQ MILES** (Figure 4-56). Click OK.

Ⓔ Scroll to the end of the table to see the new field.

It has <Null> values for all records. For more information, see the call-out box *Null values*.

Ⓕ Add another field and give it the following properties:

- Name: **POPDENSITY**
- Type: Float
- Alias: **POP DENSITY**

Ⓖ Compare your Add Field dialog box to Figure 4-57 and click OK.

Figure 4-56

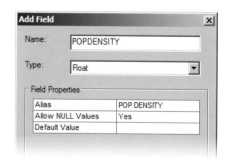

Figure 4-57

Usually, when you make changes to a layer in ArcMap, you affect only the layer and not the feature class it points to. Changing the structure of a table by adding or deleting fields is an exception.

Ⓗ In the Catalog window, under ReadyData, right-click *BlockGroups* and choose Properties.

Ⓘ In the Feature Class Properties dialog box, click the Fields tab.

You can see that both fields have been added to the feature class.

Ⓙ At the bottom of the list of fields, click the gray box to the left of the POPDENSITY field name (Figure 4-58).

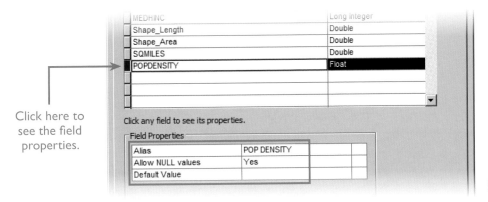

Click here to see the field properties.

Figure 4-58

The alias that you typed (POP DENSITY) has been added, too.

Ⓚ Close the Feature Class Properties dialog box.

Ⓛ In the *BlockGroups* attribute table, right-click the SQ MILES field heading and choose Calculate Geometry.

A warning message appears (Figure 4-59).

Figure 4-59

You can calculate and edit attribute values without starting an "edit session"—we're about to do it—but you give up the option of undoing a mistake. That's a problem if you overwrite data that you can't recover, but in many situations you can fix a mistake by running the calculation again.

Ⓜ On the Calculate Geometry warning, check the box that says "Don't warn me again" and click Yes.

N In the Calculate Geometry dialog box, make sure the Property is Area and the Coordinate System option is "Use coordinate system of the data source."

O Click the Units drop-down arrow and choose Square Miles US [sq mi] (Figure 4-60). Click OK.

Figure 4-60

ArcMap calculates the area in square miles for each record and adds the values to the table (Figure 4-61).

POP OVER 18	MEDIAN HH INCOME	Shape_Length	Shape_Area	SQ MILES	POP
761	29782	10067.939896	4336513.374176	0.155552	<Null>
1400	60122	9614.548051	5620129.655005	0.201595	<Null>
647	56282	7154.443203	2311745.541998	0.082923	<Null>
1501	38276	5921.398897	1661935.029169	0.059614	<Null>
1708	94370	18634.183316	17181099.509898	0.61629	<Null>
710	74676	9531.281284	2966226.487788	0.106399	<Null>
2659	42003	9436.441599	4270425.38801	0.153181	<Null>
1470	60828	8160.869766	3812504.144359	0.136755	<Null>

BlockGroups

I◄ ◄   0 ► ►I (0 out of 6421 Selected)

Figure 4-61

P In the table, right-click the POP DENSITY field heading and choose Field Calculator.

In the Field Calculator dialog box, ArcMap automatically supplies the first part of the expression: "POPDENSITY = ."

Q In the list of fields, double-click TOTPOP to add it to the expression box.

R Click the Division button *I*.

S In the list of fields, double-click SQMILES.

The complete expression should be:

POPDENSITY = [TOTPOP] / [SQMILES] (Figure 4-62).

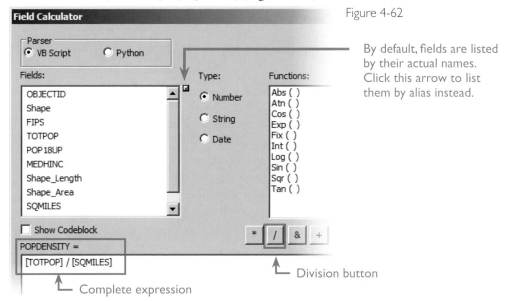

Figure 4-62

By default, fields are listed by their actual names. Click this arrow to list them by alias instead.

Complete expression

Division button

T Click OK.

ArcMap calculates the population density for each record and adds the values to the table (Figure 4-63).

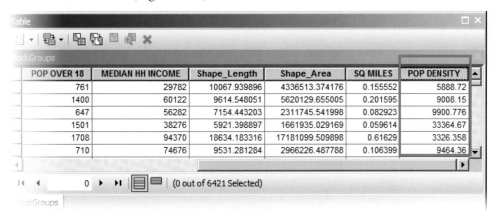

POP OVER 18	MEDIAN HH INCOME	Shape_Length	Shape_Area	SQ MILES	POP DENSITY
761	29782	10067.939896	4336513.374176	0.155552	5888.72
1400	60122	9614.548051	5620129.655005	0.201595	9008.15
647	56282	7154.443203	2311745.541998	0.082923	9900.776
1501	38276	5921.398897	1661935.029169	0.059614	33364.67
1708	94370	18634.183316	17181099.509898	0.61629	3326.358
710	74676	9531.281284	2966226.487788	0.106399	9464.36

0 out of 6421 Selected

Figure 4-63

5) Add fields to calculate the percentage under 18. To get the percentage of the population under age 18, we'll subtract the 18-and-over population from the total population and divide the result (the under-18 population) by the total population. Again, we'll need two new fields to hold our calculations.

A At the top of the table, click the Table Options button and choose Add Field.

Figure 4-64

Figure 4-65

B Give the new field these properties:
- Name: POPUNDER18
- Type: Short Integer
- Alias: POP UNDER 18

C Compare the Add Field dialog box to Figure 4-64. Click OK.

D Add another field with these properties::
- Name: PCTUNDER18
- Type: Float
- Alias: PCT UNDER 18

E Compare your Add Field dialog box to Figure 4-65 and click OK.

▷ If you realize, after adding a field, that you've made a mistake—especially if you've set the wrong field data type—right-click the field heading in the table, choose Delete Field, and repeat the operation.

The two new fields are added to the end of the table with <Null> values.

6) **Calculate the fields.** Now we'll calculate values for the fields.

A In the table, right-click the POP UNDER 18 field heading and choose Field Calculator.

B Click Clear to remove the previous expression.

C In the list of fields, double-click TOTPOP to add it to the expression box.

D Click the Minus button ⎯ .

E In the list of fields, double-click POP18UP.

The finished expression should be:

POPUNDER18 = [TOTPOP] - [POP18UP] (Figure 4-66).

F Click OK on the Field Calculator dialog box.

ArcMap calculates the population under age 18 for each record (Figure 4-67).

Figure 4-66

Figure 4-67

The last block group calculation we have to make is to divide the under-18 population by the total population to get the percentage of children. We have a potential problem with records that have a total population of 0.

Ⓖ Scroll back across the table to see the TOTAL POP field.

If we were to scroll down the records, eventually we would find at least one with a population of 0. Block groups can occasionally be uninhabited. Since division by zero is undefined, our calculations will give us an error message. Instead, we'll use an attribute query to select the records with nonzero values in the TOTAL POP field. Then we'll run the calculation on just those records.

Ⓗ At the top of the table, click the Select by Attributes button 🔳.

Ⓘ In the Select by Attributes dialog box, double-click "TOTPOP" in the list of fields to add it to the expression box.

Ⓙ Go on to make the following expression: "TOTPOP" > 0.

By default, this dialog box, like the Field Calculator, displays field names, not aliases.

Ⓚ Compare your expression to Figure 4-68 and click Apply.

Ⓛ Close the Select by Attributes dialog box.

In the table, 6,405 of 6,421 records are selected.

Ⓜ Scroll to the end of the table, right-click the PCT UNDER 18 field heading, then choose Field Calculator.

Ⓝ Click Clear to remove the previous expression.

Ⓞ In the expression box, make the following expression: ([POPUNDER18] / [TOTPOP]) * 100.

▷ Make sure to include the parentheses as shown.

Ⓟ Compare your expression to Figure 4-69 and click OK.

Values are calculated for the selected records. The unselected records still have <Null> values. We'll calculate these to zero, because if no one lives in the block group, it's reasonable to say that the percentage of children is zero.

Ⓠ At the top of the table window, click the Switch Selection button 🔁.

A message warns you that the operation might take a long time, but don't worry—it won't.

Ⓡ Click Yes to continue.

Ⓢ Right-click the PCT UNDER 18 field heading again and choose Field Calculator. Clear the expression.

Ⓣ In the expression box, type 0 and click OK.

Ⓤ At the top of the table, click the Clear Selection button 🔲.

Figure 4-68

Figure 4-69

Type the parentheses from the keyboard. Spaces between parentheses and field name brackets are optional.

We now have the attributes we need for population density and percentage of the population under age 18 (Figure 4-70).

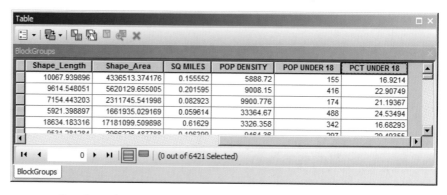

Figure 4-70

ⓥ Close the table window.

ⓦ Save the map.

ⓧ If you're continuing to the next exercise now, leave ArcMap open; otherwise, exit ArcMap.

Null values

When a field is added to a geodatabase table, it's populated with <Null> values. A <Null> value indicates the absence of data and usually means that data hasn't been obtained or that values are unknown for some reason. <Null> values are different from zeroes and blank text strings. The query "FIELDNAME" IS NULL will select records with <Null> values. It will not select records with blank text strings or zero values.

An important difference between the geodatabase feature class and shapefile formats is that shapefiles don't support <Null> values. When you add a text field to a shapefile, it's populated with blank values. When you add a numeric field, it's populated with zeroes. If you convert the shapefile to a geodatabase feature class, the blank strings and zeroes will be preserved, which may cause confusion. In shapefiles, arbitrary codes, such as -9999, are often used to represent a <Null> value. Using -9999 may cause serious problems, like when calculating summary statistics.

Exercise 4h: Prepare the parcel data

Our search for suitable park sites is limited to vacant land parcels. As we saw in Lesson 2, vacancy is an attribute of the stand-alone, nonspatial *Vacant Parcels* table. We need to associate this attribute with the *Parcels* feature class so that we can see and select vacant parcels on the map. We'll do this with an operation called a table join. A table join attaches the attributes of one table (usually, a nonspatial stand-alone table) to those of another (usually, a feature class attribute table). The join is based on a field of values, typically an identification code, that is common to both tables. Values in the common field are used to match records in the stand-alone table with records in the feature attribute table. For more information, see the topic *Table joins and relates*.

1) **Add data.** First, we'll add *Parcels* and *Vacant Parcels* to the map document.

Ⓐ If necessary, start ArcMap and open Lesson4 from the list of recent maps.

Ⓑ In the Table of Contents, turn off all the layers except *LARiver* and *Counties* to simplify the map display.

Ⓒ In the Catalog window, in the SourceData folder, under City of LA, add *Parcels.shp* and *Vacant Parcels.dbf* to the map.

In the Table of Contents, the view changes to List By Source 🖳 so you can see that there is a table in the map document.

Ⓓ Zoom to the *Parcels* layer.

2) **Identify the common attribute.** We'll look at both tables to find the common attribute, or key field, to join on. Actually, when we viewed the *Vacant Parcels* item description in Lesson 2, we saw that the PIN field serves this purpose, but we should confirm that.

Ⓐ Open the *Parcels* attribute table, and then the *Vacant Parcels* table.

Ⓑ At the bottom of the table window, drag and drop the Vacant Parcels tab onto the blue arrow pointing to the right (Figure 4-71).

▷ You don't see the arrows until you start to drag the tab.

Ⓒ In the *Parcels* table (on the left side of the table window), scroll all the way to the right to see the PIN attribute.

▷ Optionally, enlarge the table window by dragging a side or corner.

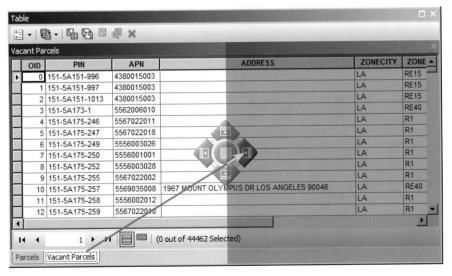

To see both tables at the same time, drag a table by its tab and drop it on a blue arrow.

Figure 4-71

Table joins and relates

Tables can be associated in ArcMap through either a join or a relate. A join attaches one table's attributes to the other table. (The join is "virtual," meaning that it exists only in the map document; the two tables remain separate on disk.) A relate is a look-up relationship, in which you select records in one table to see matching records in the other table. A join is stronger than a relate because attributes in a joined table can be used to symbolize and query features on the map, while attributes from a related table cannot. Both joins and relates require a common attribute to match records.

A join is appropriate when, for any given record in the feature class table (the table *to* which attributes are being attached), there is at most one possible match in the stand-alone table (the table *from* which attributes are being attached). A relate is appropriate when more than one match is possible. Consider some examples:

1) A feature class of parcels and a stand-alone table of vacant parcels, where the common attribute is PARCEL ID, a unique identifier in both tables. Each record in the *Parcels* table has at most one matching record in the *Vacant Parcels* table, so a join may be used. The records have a one-to-one relationship: any given record in either table can have no more than one match in the other table. (Not all records in the *Parcels* table will have matches because not all parcels are vacant.).

2) A feature class of rivers and a stand-alone table of watersheds, where the common attribute is WATERSHED ID. This is a nonunique identifier for rivers, because all rivers that drain into the same watershed, or drainage area, have the same WATERSHED ID. It is a unique identifier for watersheds. Because a river drains into just one watershed, each record in the *Rivers* table has at most one matching record in the *Watersheds* table, which means that a join may be used. The records have a many-to-one relationship: many records in the feature class table can match the same record in the stand-alone table.

The same watershed record, such as Los Angeles River Watershed, may match many river features.

Attribute from *Rivers* feature class   Key field   Attribute from *Watersheds* table

3) A feature class of the Los Angeles River and a stand-alone table of improvement projects, where the common attribute is RIVER ID. This is a unique identifier for river features. It is a nonunique identifier for the stand-alone table, because more than one improvement project may be located along the same river. In this situation, a relate should be used because a given record in the feature class table may have more than one match in the stand-alone table. In a join, only one matching record could be attached to the feature class table; the others would be left out. The records have a one-to-many relationship: one record in the feature class table may have many matches in the stand-alone table.

Select a feature in the *Rivers* feature class table and click the Related Tables button to see matching records in the stand-alone table.

D In the *Vacant Parcels* table (on the right), make sure you can also see the PIN attribute (Figure 4-72).

Figure 4-72

The PIN, or Parcel Identification Number, is a unique identifier. In theory, when we join the tables, each of the 44,462 records in the *Vacant Parcels* table should find a matching record in the *Parcels* table. In reality, with this many records, a perfect result is unlikely.

3) **Join tables.** We don't need to have the tables open to join them.

A Close the table window.

B In the Table of Contents, right-click the *Parcels* layer and choose Joins and Relates → Join.

C In the Join Data dialog box, make sure that the drop-down list at the top is set to "Join attributes from a table."

D Set drop-down list number 1 to PIN.

This specifies the key field in the *Parcels* table.

E Confirm that drop-down list number 2 is set to *Vacant Parcels.*

ArcMap has defaulted to this table as the join table.

F Make sure that drop-down list number 3 is set to PIN.

This is the key field in the *Vacant Parcels* table. ArcMap defaults to it because it's the same field name. (Key fields don't have to have the same name, however—just a compatible data type.)

G In the Join Options, click "Keep only matching records."

The only parcels that interest us are the vacant ones—those with matches in the *Vacant Parcels* table.

H Compare your dialog box to Figure 4-73.

▷ Don't click OK yet.

Figure 4-73

I Click Validate Join.

As the join is validated, you're prompted to create an attribute index on the PIN field in the *Vacant Parcels* table (Figure 4-74).

Figure 4-74

J On the Create Index prompt, click Yes.

It takes a while to validate the join. When the process is done, you'll get a message (Figure 4-75).

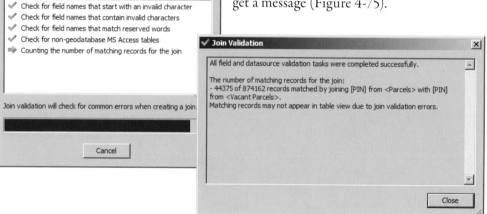

Figure 4-75

K On the Join Validation message, click Close.

L On the Join Data dialog box, click OK.

Attribute indexes

Creating an attribute index may improve the speed of the join operation—depending on the size of the table being joined—and is generally a good idea. An index also improves performance when the indexed field is queried or used to symbolize features.

Indexed fields are marked with an asterisk in attribute tables. To add or remove an attribute index from a field, open the feature class or table properties and click the Indexes tab. (The OBJECTID field in geodatabase feature classes is indexed automatically.)

When the operation is done, the *Parcels* layer redraws in the map. Because you chose to keep only the matching records, vacant parcels are the only ones that draw (Figure 4-76).

Ⓜ Open the attribute table for the *Parcels* layer.

Ⓝ At the bottom of the table, click the Move to end of table button ▶| .

Ⓞ Scroll across the attributes.

The joined table has the attributes of both tables. It has 44,375 records, which means that 87 records from the *Vacant Parcels* table were unmatched (Figure 4-77). That's a match rate of 99.8 percent.

The asterisk on the PIN field means that it's indexed.

AREA	PERIMETER	OID	PIN *	APN	ADDRESS
12101.052224	820.780539	44387	036B213-678	7426029913	
11093.190397	1065.349607	44392	036B213-679	7426029005	
2979.246067	270.897543	44365	138B141-1108	4429020024	
1773.117778	357.149721	44367	159B169-2212	2377015021	
12695.3635	664.423561	44383	138B121-265	4431040017	
7077.468121	357.726029	44373	189B161-1201	2304015034	
7076.564826	357.706328	44391	189B161-1200	2304015033	
7112.516589	360.04589	44386	189B161-1202	2304015035	
7563.828422	378.088101	44374	189B161-1203	2304015036	
8416.484943	399.00321	44384	189B161-1204	2304015037	
12705.613707	450.998044	44397	132B189-1260	5081025024	960 S WILTON PL LOS ANGELES 90019
3007.57661	271.382648	44412	162B149-698	2277025007	
6695.761639	339.730503	44408	195B173-642	2408006014	

44375 (0 out of 44375 Selected)

Figure 4-76

Figure 4-77

Ⓟ Close the table.

4) **Copy features.** We'll create a new feature class in ReadyData, copying just the vacant parcel features from the joined table.

Ⓐ Open the Copy Features tool.

▷ How? Open the Search window and click the Copy Features tool. (If the tool isn't in the list, type its name in the search box.)

Ⓑ In the Copy Features dialog box, set the Input Features to the *Parcels* layer.

Ⓒ In the Output Feature Class parameter, clear the path, type **VacantParcels**, then press Tab.

Ⓓ Compare your dialog box to Figure 4-78 and click OK.

Copy Features

Input Features
Parcels

Output Feature Class
C:\UGIS\ParkSite\AnalysisData\ReadyData.gdb\VacantParcels

Configuration Keyword (optional)

Output Spatial Grid 1 (optional)

Figure 4-78

When the tool is finished, the new feature class appears in ReadyData and a layer is added to the map.

E In the Table of Contents, remove the (old) *Parcels* layer and the (old) *Vacant Parcels* table.

F At the top of the Table of Contents, click the List By Drawing Order button 🖳.

G Open the attribute table for the (new) *VacantParcels* layer and scroll across the attributes.

The virtual attributes of the joined table are now permanent attributes of the new feature class.

Field Properties

Name: Vacant_Parcels_ZONECITY

Alias: ZONECITY

Type: String

Display

Figure 4-79

H Right-click any field heading in the table (other than a software-managed field) and choose Properties.

Field names from the two original tables in the join have been converted to aliases in this table. The field names now indicate which source table the attribute came from (Figure 4-79).

I Close the Field Properties dialog box.

J Close the table.

5) Dissolve vacant parcels. We still need to calculate acreage for the vacant parcels (Row 3 of the requirements table). At this point we might have a new thought: what if there are some adjacent vacant parcels that are smaller than an acre individually but would be more than an acre if combined? We might be missing out on some potential park sites simply because of property lines (Figure 4-80).

The example shows three adjacent vacant parcels. Separately, each is less than an acre, but together they're more than an acre. We want to combine them into a potential park site.

.25 acres

.71 acres

.33 acres

Figure 4-80

To address this possibility, we'll dissolve the *VacantParcels* feature class, which will create single features out of parcels sharing a common boundary. In the process, we're going to lose attributes. We don't really need the attributes though—at least, we don't need them as much as we need good park locations.

A From the ArcMap main menu, choose Geoprocessing → Dissolve.

B In the Dissolve dialog box, set the Input Features to the *VacantParcels* layer.

❸ Change the output feature class name to **VacantParcels2**.

❹ In the Dissolve Fields area, leave all the fields unchecked.

We're dissolving purely on the parcel geometry. See the topic *Dissolving features* for a discussion on dissolving on geometry and dissolving by attributes.

❺ At the bottom of the dialog box, uncheck the Create multipart features box.

If we allowed multipart features, the output feature class would consist of a single multipart polygon. We'd have one great big discontinuous vacant parcel.

❻ Compare your dialog box to Figure 4-81 and click OK.

When the dissolve is complete, *VacantParcels2* is added to ReadyData and a layer is added to the map.

❼ In the Table of Contents, turn *VacantParcels2* off and on a few times.

Even with the map zoomed out to the city limits, you can see some areas where parcel boundaries have disappeared.

❽ In the Table of Contents, remove the (old) *VacantParcels* layer.

❾ Open the attribute table for the (new) *VacantParcels2* layer.

It has 15,802 records, about a third as many as the undissolved layer. The only attributes are the four managed by ArcMap (Figure 4-82).

Uncheck this box

Figure 4-81

Figure 4-82

Figure 4-83

6) Calculate acreage. Let's get back to acreage. ArcMap has recalculated the Shape_Area field for the dissolved parcels, but the units are in square feet. We need to convert them to acres.

Ⓐ Add a new field to the *VacantParcels2* table.

▷ How? Click the Table Options button at the top of the table window and choose Add Field.

Ⓑ Give the new field these properties:

- Name: ACRES
- Type: Double

Ⓒ Compare your Add Field dialog box to Figure 4-83 and click OK.

Ⓓ In the table, right-click the ACRES field heading and choose Calculate Geometry.

Ⓔ In the Calculate Geometry dialog box, make sure the Property is Area and the coordinate system option is "Use coordinate system of the data source."

Ⓕ Click the Units drop-down arrow and choose Acres US [ac].

Ⓖ Compare your Calculate Geometry dialog box to Figure 4-84 and click OK.

ArcMap calculates the acreage for each record and adds the values to the table (Figure 4-85).

Ⓗ Close the table.

Figure 4-84

Figure 4-85

7) Clean up the ReadyData geodatabase. We'll get rid of the undissolved *VacantParcels* feature class, and rename the dissolved one. Then we'll edit the item description for this feature class.

Ⓐ In the Catalog window, expand ReadyData.gdb.

Ⓑ Right-click *VacantParcels* and choose Delete.

Make sure *VacantParcels*—not *VacantParcels2*—is highlighted in the Catalog window.

Ⓒ Click Yes on the prompt to confirm your deletion.

The feature class is deleted from the geodatabase.

Ⓓ In the Catalog window, right-click *VacantParcels2* and choose Rename.

Ⓔ Rename the feature class **VacantParcels** and press Enter.

Ⓕ Right-click the renamed *VacantParcels* feature class and choose Item Description (Figure 4-86).

Figure 4-86

Except for the title (which reflects the old feature class name) and the file format at the top, the Item Description is empty.

Ⓖ In the Item Description window, click the Preview tab.

Ⓗ In the row of tools at the top of the window, click the Create Thumbnail button 🔲.

Ⓘ Click the Description tab.

The Item Description is updated with a thumbnail graphic.

Ⓙ On the Description tab, click Edit.

Ⓚ In the Title box, change the title to **Vacant Parcels**.

Ⓛ Scroll down and click in the Tags box. Type the following: **parcels, vacant parcels, Los Angeles** (Figure 4-87).

▷ Include the commas: they separate tags.

Figure 4-87

Figure 4-88

Figure 4-89

M Scroll down, if necessary. Click in the summary box and type **This dataset of land parcels in the city of Los Angeles supports queries on the size of contiguous vacant parcels.**

N Click in the Description box and type **This dataset represents 44,375 original vacant parcels in the city of Los Angeles. Adjacent parcels have been dissolved, leaving 15,802 features.**

O Click in the Credits box and type: **City of Los Angeles**.

P At the top of the Description tab, click Save.

The Item Description should look like Figure 4-88.

Q Close the Item Description.

R Optionally, in the Table of Contents, rename the *VacantParcels2* layer to **VacantParcels**.

S In the Catalog window, confirm that your ReadyData geodatabase looks like Figure 4-89.

T Save the map document.

U Close the data requirements table, if you have it open.

V Exit ArcMap.

The project geodatabase has been created and the data has been prepared for the analysis. In the next lesson, we'll make edits to park features and attributes.

Deleting items from the catalog

Items you delete from ArcCatalog, or from the Catalog window in ArcMap, are permanently removed from your computer. They cannot be retrieved from the Recycle Bin in Windows or undeleted by any ArcGIS command. This is true for any item shown in the catalog: datasets, toolboxes, map documents, and files of all other kinds. Be careful about deletions and always keep backup copies of your work. In the Catalog window, you can easily copy an item (including a geodatabase or a folder) by right-clicking it and choosing Copy.

Lesson

5 Edit data

DATA INTEGRITY IS CRUCIAL IN GIS,

so it shouldn't be easy to change feature shapes or locations by accident. In ArcMap, before you can make a spatial change to a feature, you have to start an edit session. That's the first safeguard the software applies to make sure you really mean to do what you're doing. Within an edit session, if you make a mistake, you can recover with the Undo button, which lets you undo edits one by one in reverse order. If the mistakes are too complicated to resolve, you can stop an edit session without saving edits.

Ideally, data would come to us free of errors or inconsistencies, but that seldom happens. Careful exploration of data usually reveals imperfections, which arise for all kinds of reasons. Sometimes data is captured or created incorrectly to begin with. Sometimes errors creep in with subsequent data processing. Sometimes the data is perfectly good until the world changes. And sometimes the data isn't wrong at all: it's just generalized for use at a certain scale and inappropriate at other scales. For example, the parks data we're using was really designed for medium-scale maps, more or less in the range of 1:100,000 to 1:250,000. When the features are examined at much larger scales, it's not surprising that they don't conform to high-resolution imagery.

ArcMap has many tools for editing features in specialized ways, as well as methods for evaluating and maintaining data integrity. Data editing, however, isn't the focus of our project. We're correcting one park feature and creating another because, during our data exploration, we saw a need to do so. Then we'll move ahead with the analysis.

Lesson Five roadmap

You are here:

What you'll do in this lesson:

1 Frame problem and explore study area ✓

2 Preview data ✓

3 Choose the data ✓

4 Build the database ✓

5 Edit data

6 Conduct the analysis

7 Automate the analysis

8 Present analysis results

9 Share results online

Edit a feature 5a

Create a new park feature 5b

Exercise 5a: Edit a feature

In Lesson 3, we saw that Pecan Playground, a park near Dodger Stadium, didn't match up very well against imagery. In this exercise, we'll edit the park boundary and update its acreage attribute.

1) **Add data to ArcMap.** We need the parks data and a basemap of imagery.

Ⓐ Start ArcMap and open a blank map.

Ⓑ On the Standard toolbar, click the Add Data drop-down arrow and choose Add Basemap.

Ⓒ In the Add Basemap dialog box, click the Imagery basemap and click Add.

Ⓓ In the Catalog window, under Folder Connections, navigate to C:\UGIS\ParkSite\AnalysisData\ReadyData.gdb.

  ▷ If your ReadyData geodatabase isn't up-to-date, download the results for Lesson 4 from the Understanding GIS web site.

Ⓔ From ReadyData, drag and drop *Parks* onto the map.

  ▷ If you get a Geographic Coordinate Systems Warning, click Close.

Ⓕ Right-click the *Parks* layer and choose Zoom To Layer.

Ⓖ Optionally, change the symbol for the *Parks* layer to a shade of green (Figure 5-1).

Figure 5-1

In a little while, we'll use My Places to go to Pecan Playground. If you removed the Data Frame Tools toolbar, you need to add it.

Ⓗ If necessary, from the main menu, choose Customize ➔ Toolbars ➔ Data Frame Tools.

2) **Start an edit session.** Most basic editing tools are found on the Editor toolbar.

Ⓐ On the Standard toolbar, click the Editor Toolbar button ⌖ to open the Editor toolbar (Figure 5-2).

You can also open the Editor toolbar in the usual way from the Customize menu.

Figure 5-2

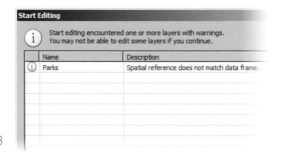

Ⓑ Place the toolbar wherever you want.

The tools are disabled because you haven't started an edit session.

Ⓒ On the Editor toolbar, click the Editor drop-down menu and choose Start Editing.

Immediately, you encounter a warning (Figure 5-3).

Figure 5-3

The warning tells you that the spatial reference (the coordinate system) of the *Parks* layer doesn't match the data frame. Recall from Lesson 3 that the first layer added to the data frame sets the data frame's coordinate system. We added the *Basemap* layer first. Because its coordinate system is WGS 1984 Web Mercator, this is also the coordinate system of the data frame. The *Parks* layer has been projected to Web Mercator on the fly, but its native coordinate system is California State Plane Zone 5. That explains the mismatch.

Is this a problem? Generally speaking, operations performed on data which is projected on the fly—whether spatial query, analysis, editing, or anything else—are safe. Because data editing is a highly sensitive operation, however, the best practice is to edit data while it is displayed in its native coordinate system, not projected on the fly. We'll take this opportunity to change the data frame's coordinate system to match the *Parks* layer.

Ⓓ On the Start Editing warning, click Stop Editing.

3) **Change the data frame's coordinate system.** This operation should be familiar to you from Lesson 3.

Ⓐ Open the data frame properties.
  ▷ How? At the top of the Table of Contents, double-click the data frame name (Layers).

B In the Data Frame Properties dialog box, click the Coordinate System tab.

C Scroll to the bottom of the list of coordinate systems and expand the Layers folder.

This folder lists the coordinate systems associated with layers in the data frame.

D Expand both coordinate systems to see which layer is associated with which system.

Figure 5-4

As you already knew, the *Parks* layer we want to edit is in the State Plane system. The *Basemap* layer—which is actually a group layer composed of multiple imagery layers—is in the Web Mercator system.

E Click the NAD_1983_StatePlane_California_V_FIPS_0405_ Feet coordinate system to select it (Figure 5-4).

Applying this coordinate system to the data frame will trigger another geographic coordinate system warning (because now the *Basemap* layer won't match). To avoid that, we'll change the *Basemap* layer's geographic coordinate system from WGS84 to NAD83. We did the same transformations in Lesson 4: the difference this time is that the transformation will be applied only within the data frame; the source datasets themselves will not be changed in any way.

Click this coordinate system.

F On the lower left side of the dialog box, click Transformations.

G In the Geographic Coordinate System Transformations dialog box, confirm these settings (or make them if necessary):

• In the Convert from box, GCS_WGS_1984 is selected.

• The Into box is set to GCS_North_American_1983.

H In the Using box, click the drop-down arrow and choose WGS_1984_(ITRF00)_ To_NAD_1983. (It should be at the top.)

I Confirm that your settings match (Figure 5-5) and click OK.

J Click OK on the Data Frame Properties dialog box.

You don't get a geographic coordinate systems warning because everything is reconciled within the data frame.

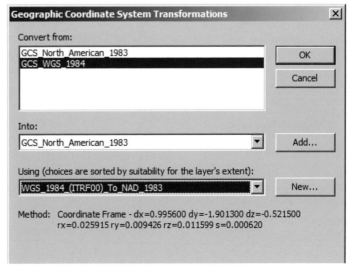

Figure 5-5

4) **Zoom to Pecan Playground.** Let's zoom in to the feature we want to edit.

Ⓐ Optionally, maximize the ArcMap window.

Ⓑ On the Data Frame Tools toolbar, click the My Places button 🚩 (Figure 5-6).

Figure 5-6

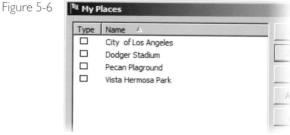

Ⓒ In the My Places dialog box, right-click Pecan Playground and choose Zoom To.

▷ If you don't have this place, download the Lesson 3 results from the Understanding GIS web site. Extract the contents and copy the *Places.dat* file to your MapsAndMore folder. Then, in the My Places dialog box, click the Load button and open this file.

Ⓓ Close My Places.

We should change the park symbol so we can see what we're doing while we're editing.

Ⓔ Open the Symbol Selector for the *Parks* layer and change the fill color to No Color.

Ⓕ Change the outline color to Quetzel Green (Figure 5-7) and the outline width to 2. Click OK on the Symbol Selector.

5) **Edit the Pecan Playground feature.** We're ready to start editing. Don't worry too much about your results—it should be fairly easy to make the feature better than it is now, and that's good enough.

Ⓐ On the Editor toolbar, click the Editor drop-down menu and choose Start Editing.

Several tools on the Editor toolbar become enabled. On the right side of the application, a Create Features window opens (Figure 5-8). (This window opens by default only in your very first ArcMap editing session; if you happen to have opened and closed it in any previous ArcMap session, you won't see it now.)

At the moment, we don't want to create new park features—we'll do that in the next exercise—we just want to edit an existing feature.

Ⓑ If necessary, close the Create Features Window.

Figure 5-7

Figure 5-8

ⓒ Zoom and pan as needed so you see the whole block, as in Figure 5-9.

The feature as drawn doesn't include the swimming pool or the play area at the north end of the block. Also, its boundaries extend into the street. (The southwest corner of the block is correctly excluded: this is a school.) After editing, Pecan Playground should look more like Figure 5-10.

Swimming pool   Playground

Figure 5-9

Basemap layers are updated periodically, so your imagery may look different.

Figure 5-10

ⓓ On the Editor toolbar, if necessary, click the Edit tool ▸.

ⓔ Click anywhere inside the Pecan Playground feature to select it.

The feature outline turns blue.

ⓕ On the Editor toolbar, click the Edit Vertices button ⬚.

Two things happen. One is that a new toolbar, the Edit Vertices toolbar, is added (Figure 5-11).

Figure 5-11

The other is that the feature is now marked with a number of small green squares (and one red one). This representation of the feature is called its "edit sketch." The squares, or vertices, are coordinate pairs that define the feature's shape and position. The red vertex is the last one added when the feature was created or edited. In Figure 5-12, though not on your screen, the vertices are numbered for reference. For more information, see the topic *Edit sketches and snapping*.

Figure 5-12

As you edit the feature, you may find it useful to zoom and pan with shortcut keys.

Important editing shortcut keys

To navigate while editing vertices, use these shortcut keys:

Z to zoom in; X to zoom out; C to pan

B for zoom/pan (click and drag to zoom; right-click and drag to pan)

Clicking navigation tools (or other tools) on the Standard toolbar interrupts vertex editing. If you accidentally interrupt vertex editing, click the Edit Vertices button on the Editor toolbar to resume it.

G On the Edit Vertices toolbar, make sure the Modify Sketch Vertices tool ▷ is selected.

H Place the mouse pointer over Vertex 2, which is in a street intersection.

When the mouse pointer is over the vertex, it changes to a four-headed arrow.

I Click and drag the vertex to where it should go—the southeastern corner of the soccer field—then release the mouse button (Figure 5-13).

▷ If you make a mistake, you can recover with the Undo button ↶ on the Standard toolbar.

Figure 5-13

Put the mouse pointer over the vertex you want to edit

Click and drag to the desired location

Release the mouse button

The blue highlight continues to show the original shape of the feature.

J Place the mouse pointer over Vertex 0 and move the vertex to the northeastern corner of the park (Figure 5-14).

The eastern boundary now has a kink because of Vertex 1. This vertex is superfluous.

K On the Edit Vertices toolbar, click the Delete Vertex tool ▷.

L Drag a small box around Vertex 1 (Figure 5-15) and release the mouse button to delete the vertex.

Figure 5-14

Figure 5-15

The eastern side of the park should now be straight, as in Figure 5-16.

Figure 5-16

Figure 5-17

Figure 5-18

Ⓜ On the Edit Vertices toolbar, click the Modify Sketch Vertices tool ▷ and move Vertex 7 (the red vertex) to the northwest corner of the park (Figure 5-17).

Ⓝ Move Vertex 5 to where the baseball diamond meets the schoolyard (Figure 5-18).

Ⓞ Click the Delete Vertex tool ▷ and delete Vertex 6 to straighten the western side of the park (Figure 5-19).

Ⓟ Click the Modify Sketch Vertices tool ▷ and move Vertex 3 to the southwestern corner of the soccer field (Figure 5-20).

Figure 5-19

Figure 5-20

6) **Explore snapping.** When you move a vertex near another feature, the vertex connects to it automatically. This behavior, called snapping, helps prevent errors such as small gaps and overshoots. A vertex will snap to features in its own layer, including its own edit sketch, as well as to features in other layers. You may have seen this behavior already with one or two of the vertices you've moved.

Ⓐ Hover over Vertex 4, the only one you haven't edited yet. (It's at the northeastern corner of the school.)

Figure 5-21

This vertex is already at a good location. We're just going to explore snapping behavior a little bit.

Ⓑ Without releasing the mouse button, drag the vertex a small distance in various directions.

As you move south or west along the edges of the feature sketch, you'll see the snap tip "Parks: Edge" appear next to the mouse pointer (Figure 5-21). The snap tip tells you that you're within the range, or snapping tolerance, of an edge in the *Parks* layer. As you move back toward your original position, you'll see the snap tip "Parks: Vertex." Moving in any direction away from the feature, you'll see no snap tip.

Ⓒ Drag the vertex back towards its original position. When you see the "Parks: Vertex" snap tip, release the mouse button.

Ⓓ Hold down the Z key on the keyboard and zoom in on Vertex 4.

▷ If you zoom in too far and the imagery disappears, hold down the X key and zoom out a little bit.

Ⓔ Place the mouse pointer over Vertex 4 again. Without releasing the mouse button, drag the vertex a small distance in different directions.

Figure 5-22

The snapping behavior is the same as before, but the snapping tolerance—which is ten screen pixels by default—now corresponds to a much smaller ground distance. That gives you more control over the exact placement of the vertex.

Ⓕ Again, drag the vertex back towards its original position. When you see the "Parks: Vertex" snap tip, release the mouse button.

Ⓖ Zoom out to see the entire edit sketch (Figure 5-22).

▷ You can use the X key on your keyboard, the Go Back To Previous Extent button ⬅, or My Places. If you need to pan, use the C key on your keyboard.

7) **Save edits.** We'll finish the sketch and save the edits.

Ⓐ On the Edit Vertices toolbar, click the Finish Sketch button 🖵.

Figure 5-23

The edited park is now a selected feature (Figure 5-23). We could still undo these edits, or stop the edit session without saving them, but we should be pretty satisfied with the new park boundary.

Ⓑ On the Editor toolbar, click the Editor drop-down arrow and choose Save Edits.

8) Update attribute values. Changing the park's shape has changed its area. That means the ACRES attribute value is no longer correct.

Ⓐ Open the *Parks* attribute table.

Ⓑ At the bottom of the table, click the Show selected records button ▤. Scroll across to see the geometry attributes.

ArcMap has automatically updated the SHAPE_LENGTH and SHAPE_AREA fields, but it's our job to maintain the ACRES attribute. The value stored now is just over four acres. This is the acreage of the old, preedited feature.

Ⓒ Right-click the ACRES field heading and choose Calculate Geometry.

Ⓓ Make sure the Property is set to Area and the Coordinate System is set to the data source.

Ⓔ Click the Units drop-down arrow and choose Acres US [ac].

Ⓕ Compare your dialog box to Figure 5-24 and click OK.

The new value is written to the table. It shouldn't be much different from the old one. Although the park was enlarged on its northern boundary, it was narrowed along three sides.

Ⓖ On the Editor toolbar, click the Editor drop-down arrow and choose Save Edits.

Ⓗ At the bottom of the table, click the Show all records button ▥.

Ⓘ At the top of the table, click the Clear Selection button ▨.

Ⓙ Close the attribute table.

Ⓚ On the Editor toolbar, click the Editor drop-down arrow and choose Stop Editing.

9) **Save the map document.** Your edits have been saved to the *Parks* feature class and are preserved whether or not you save the map document. We'll save the map document because we'll keep using it in Exercise 5b and you may not be continuing immediately.

Ⓐ Save the map document as **Lesson5** in your MapsAndMore folder.

Ⓑ If you are continuing to the next exercise now, leave ArcMap open; otherwise, exit ArcMap.

Figure 5-24

Edit sketches and snapping

When you edit a feature, you work with its edit sketch. An edit sketch is a representation of the feature with its coordinate anatomy exposed. It is made up of vertices, edges, and ends. Vertices define the shape and location of a feature, and usually mark a change in angle of the feature's shape. Edges, also called segments, are lines that connect vertices. Ends are the vertices at either end of a line feature. The special vocabulary makes it clear that vertices, edges, and ends are not features in and of themselves, but rather components of features.

A park feature

Its edit sketch

Each vertex is an x,y coordinate pair.

Every vertex is an x,y coordinate pair. You can view (and edit) the list of coordinate pairs by clicking the Sketch Properties button on the Editor Toolbar.

Vertices and edges can be moved with the mouse, or moved and edited in other ways with context menu commands.

As you move a vertex or edge, it will snap (connect automatically) to parts of nearby features. Snapping behavior is controlled with the Snapping toolbar. (On the Editor toolbar, click the Editor menu and choose Snapping → Snapping Toolbar.) By default, a vertex or edge that you're moving will snap to edges, vertices, ends, or point features in any feature-based layer in the data frame. To turn snapping off and on for any of these elements, click its button on the toolbar.

Moving an edge

Moving a vertex

to edges
to vertices
to ends
to points
(features)

The Snapping drop-down menu on the toolbar lets you turn all snapping off. It also lets you control snapping to some additional elements.

The Snapping Options dialog box lets you change the snapping tolerance from the default value of ten pixels. It also allows you to enable snapping for basemap and feature service layers, and to control the display and formatting of snap tips.

Exercise 5b: Create a new park feature

Another local park, Vista Hermosa Park, was represented by a raster dataset in our source data, but not by a feature in any of the parks feature classes. In this exercise, we'll create it by tracing its outline against the basemap image. The process is very similar to the editing that you did in the last exercise.

After creating the feature, we'll add attribute values for it. We'll also edit the attribute values for the two park features we loaded in Lesson 4.

1) **Start an edit session.** We'll zoom to the location of the missing park and start an edit session.

Ⓐ If necessary, start ArcMap and open Lesson5 from the list of recent maps.

Ⓑ Use My Places to zoom to Vista Hermosa Park and close My Places.

The park itself doesn't take up the entire L-shaped lot, just the portion shown by the dotted outline in Figure 5-25.

Figure 5-25

You won't see the dotted line on your map—it was added here for reference.

⊙ Zoom in to an area more or less matching Figure 5-25.

⊙ In the Table of Contents, right-click the *Parks* layer and choose Edit Features → Start Editing.

This is another way to start an edit session.

⊙ On the Editor toolbar, click the Create Features button ⊞.

⊙ In the Create Features window, click the Parks symbol (Figure 5-26).

▷ It may appear to be selected already, but you still need to click it.

Figure 5-26

By clicking the symbol, you've chosen a template. A template determines the properties that new features you create will have. Because our map only has one editable layer, Parks is the only template available. And because all our parks are symbolized the same way, the template has just one symbol.

⊙ Right-click the Parks template and choose Properties.

The Template Properties dialog box lets you view and change the properties of the template (Figure 5-27).

When you create a new feature in the layer, this drawing tool (or construction tool, as it is called), will be selected by default.

You can set default attribute values, such as State Park, for all new features.

Figure 5-27

We don't need to change any of these settings.

⊙ Close the Template Properties dialog box.

⊙ Close the Create Features window.

Feature templates are especially useful when you're creating many new features in a layer with varied symbology. For example, a Points of Interest layer might have different symbols and descriptive attributes for restaurants, museums, parks, and so on. A template allows you to choose the right symbol and description for each new point as you add it.

2) Create the park feature. We're ready to draw the boundary of Vista Hermosa Park. Again, our purposes don't require extremely high accuracy. We just want a boundary that approximately conforms to the yellow dotted outline in Figure 5-25. If you decide to zoom in and work at large scale, you may be able to see and follow the fence line that encloses the park. (Remember to use the editing shortcut keys.)

Ⓐ Move the mouse pointer over the map.

The cursor becomes a crosshairs.

Ⓑ Choose a beginning point.

This can be anywhere, but corners are usually good—for example, you might use the southeast corner of the soccer field.

Ⓒ Click once to start drawing the feature.

A red vertex is added at the location where you clicked.

Ⓓ Move the mouse in the direction you want to go. (Don't drag, just move.)

As you move, the cursor remains connected to the vertex by a purple line (Figure 5-28). As long as you're moving in a straight line, you don't need to add a vertex.

Figure 5-28

E When you come to a place where you need to change direction, click to add a second vertex.

The new vertex is red (it's the last one added) and the first vertex turns green.

F Press the Tab key.

The Feature Construction toolbar opens (Figure 5-29).

This toolbar contains specialized drawing tools that you don't really need here but that you might find useful on other occasions. When you move the pointer away from the toolbar, it fades to a semi-transparent state; it disappears when you add the next vertex to your sketch (or press the Tab key again).

G Trace the boundary of the park as well as you can, following fence lines, property lines, and sidewalks.

  ▷ If you add a bad vertex, click the Undo Add Vertex button ↺ on the Feature Construction toolbar or the Undo button on the Standard toolbar.

H When the polygon is complete, click the Finish Sketch button 🗇 on the Feature Construction toolbar (or simply double-click).

Your new feature, still selected, should resemble Figure 5-30.

Again, the result doesn't have to be perfect. If you want to make some changes, however, click the Edit Vertices 🖵 button on the Editor toolbar and go ahead. If you're really dissatisfied with your feature, press the Delete key on the keyboard to get rid of it, then start over.

I When you're satisfied with the new park, on the Editor toolbar, click the Editor drop-down arrow and choose Save Edits.

Figure 5-29

The default tool draws straight line segments.

Figure 5-30

The Feature Construction toolbar

Tools on the Feature Construction toolbar let you control the properties and behavior of line segments. For example, you can draw true curves or constrain line segments to be parallel or perpendicular to existing segments. You can also right-click while drawing a line segment to open a context menu with more options. You can drag the Feature Construction toolbar to any location on your screen without interrupting your editing. If you want the toolbar to stay open while you work, from the Editor toolbar, choose Editor → Options and, on the General tab, check the Show feature construction toolbar box. (Alternatively, from the main menu, choose Customize > Toolbars > Feature Construction.)

3) Edit attribute values. We've created the Vista Hermosa Park feature, but we haven't given it attribute values.

Ⓐ On the Editor toolbar, click the Attributes button 📇.

The Attributes window opens, showing the attributes and values for this feature (Figure 5-31).

Figure 5-31

Values in software-managed fields are grayed-out and can't be edited. (Your length and area values won't match those in the figure. Your OBJECTID value may be different, too.)

Ⓑ In the Attributes window, click the <Null> value next to NAME and enter **Vista Hermosa**.

Ⓒ Click the <Null> value next to TYPE and enter **Local park or recreation area** (Figure 5-32).

Figure 5-32

Ⓓ Close the Attributes window.

Ⓔ Save your edits.

We still need to calculate the acreage for Vista Hermosa Park and for the two park features we loaded in Lesson 4 (which didn't have an acreage attribute). We could find these records by scrolling to the bottom of the *Parks* attribute table, but a more systematic way is to query for Null values in the ACRES field.

F Open the *Parks* attribute table.

G At the top of the table, click the Select By Attributes button .

H In the Select by Attributes dialog box, make a query expression as follows:

- Make sure the Method is set to Create a new selection.
- Double-click the "ACRES" field.
- Click the "Is" button (or type the word **IS**).
- Click Get Unique Values and double-click NULL.

I Confirm that your expression matches Figure 5-33 and click Apply.

J Close the Select by Attributes dialog box.

K In the table, click the Show selected records button ▤.

Three records should be selected (Figure 5-34).

L In the table, calculate the geometry for the ACRES field.

▷ How? Right-click the ACRES field heading and choose Calculate Geometry. Set the units to Acres US [ac] and accept the other defaults.

Figure 5-33

Figure 5-34

OBJECTID *	Shape *	NAME	TYPE	ACRES	Shape_Length	Sha
168	Polygon	Los Angeles	State Historic Park	<Null>	7426.876827	13
169	Polygon	Rio de Los Angeles	State Recreation Area	<Null>	5304.017669	1
170	Polygon	Vista Hermosa	Local park or recreation area	<Null>	2773.835996	4

The results should be about 32 acres for Los Angeles State Historic Park, 36 acres for Rio de Los Angeles State Recreation Area, and nine to ten acres for Vista Hermosa Park (Figure 5-35).

Figure 5-35

NAME	TYPE	ACRES
Los Angeles	State Historic Park	32.1343
Rio de Los Angeles	State Recreation Area	36.7524
Vista Hermosa	Local park or recreation area	9.57868

The TYPE values for Los Angeles and Rio de Los Angeles (the two parks we loaded) are nonstandard in our table. All other state parks have the value "State or local park or forest."

M In the table, right-click on the gray square next to the Vista Hermosa record and choose Unselect.

That leaves two selected records.

Figure 5-36

Ⓝ Right-click the TYPE field heading and choose Field Calculator.

Ⓞ In the expression box at the bottom of the Field Calculator, type **"State or local park or forest"**.

▷ The quotation marks are required for a string value.

Ⓟ Compare your dialog box to Figure 5-36 and click OK.

The values for the two selected records are updated (Figure 5-37).

NAME	TYPE	ACRES
Los Angeles	State or local park or forest	32.1343
Rio de Los Angeles	State or local park or forest	36.7524

Figure 5-37

4) **Stop editing.** We're finished with edits to the *Parks* layer.

Ⓐ Save your edits.

Ⓑ In the table, clear the selection 🔲 and show all records 🔲.

Ⓒ Stop editing and close the attribute table.

Editing layers in different workspaces

During an edit session, you can only edit layers that derive from a single workspace—that is, from the same geodatabase or the same folder. If your layers point to different workspaces, you'll be prompted to choose which workspace you want to make editable.

5) **Zoom out and save the map.** We'll take a quick look at our work from a smaller scale and save the map.

Ⓐ Resymbolize the *Parks* layer to give it a solid green fill color and a **0.5**. outline width. Change the outline color, too, if you want.

Ⓑ Open My Places.

Ⓒ In the My Places dialog box, hold down the Shift key and select both Vista Hermosa Park and Pecan Playground (Figure 5-38).

Figure 5-38

❶ Click Zoom To and close My Places.

On the map, you're zoomed to an extent that includes both parks (Figure 5-39).

Vista Hermosa Park →

← Pecan Playground

Figure 5-39

❷ Remove the Editor toolbar, the Data Frame Tools toolbar, and any other specialized toolbars that may be open.

As mentioned earlier, edits are written to the source data at the time you save them. Although we're making a practice of saving map documents, your park edits have already been saved to the feature class.

❸ Save the map and exit ArcMap.

All our data preparations are complete. In the next lesson, we'll look for suitable park sites.

Lesson

6 Conduct the analysis

NOW WE COME TO THE POINT

of our preparatory work: the analysis. Our original feature class of parcels was like an assembled 874,000-piece jigsaw puzzle covering the entire city, each piece representing an individually owned property. In Lesson 4, we cut this down to about 45,000 parcels (those that were vacant), then to 15,000 (those that were adjacent). Now we'll continue taking pieces away, removing nonsuitable parcels until we're left with just those candidates that meet the criteria established in Lesson 2:

- On a vacant land parcel one or more acres in size
- Within the LA city limits
- Within 0.75 miles of the LA River (the closer the better)
- At least 0.25 miles from the nearest park
- In a neighborhood where population density is at least 8,000 people per square mile
- Where at least 23 percent of the population is under 18
- Where median household income is $50,000 or less
- Preferably serving the most people within a quarter-mile radius

But how does this translate into a method and a workflow? How do we take these requirements and match them to appropriate geoprocessing tools in the right sequence? Planning a workflow means identifying the tasks, finding a suitable tool for each, and understanding the inputs and outputs so we can order the operations in a logical, efficient way.

Most of the tasks in our analysis—and this is typical—fall into just a few categories: distance problems (what's near what), topological problems (what crosses, touches, or contains what), attribute query problems (what has this or that value), and overlay problems (what areas are common to features in different layers). Each of these problem categories is associated with its own special set of tools.

Lesson Six roadmap

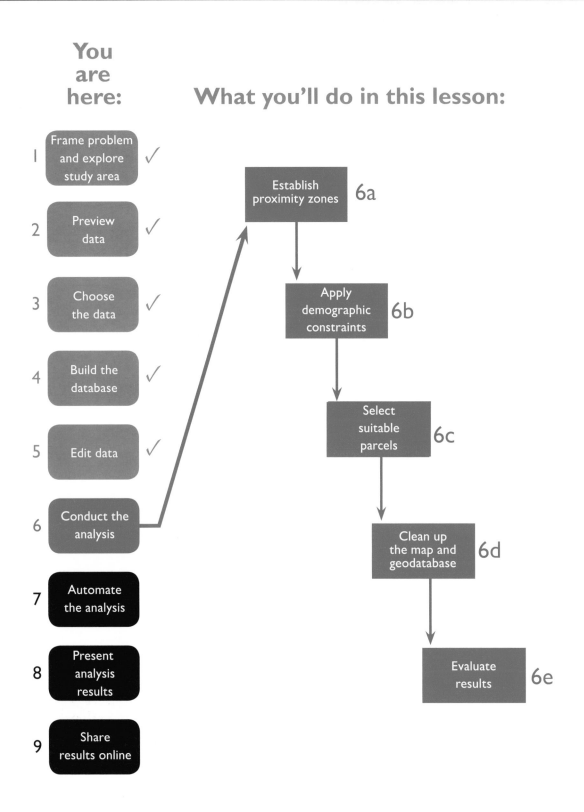

You are here:

1 Frame problem and explore study area ✓

2 Preview data ✓

3 Choose the data ✓

4 Build the database ✓

5 Edit data ✓

6 Conduct the analysis

7 Automate the analysis

8 Present analysis results

9 Share results online

What you'll do in this lesson:

Establish proximity zones — 6a

Apply demographic constraints — 6b

Select suitable parcels — 6c

Clean up the map and geodatabase — 6d

Evaluate results — 6e

Before beginning an involved analysis, you may find it helpful to draw a rough plan or diagram. Any medium will do—paper, a software application, a whiteboard—as long as you're prepared to make revisions as you go. A typical working plan (the one we'll follow) could be sketched out to look something like Figure 6-1. This sketch depicts the broad general approach, although there are bound to be a few additional twists and turns along the way.

Figure 6-1

ANALYSIS PLAN

START WITH LA RIVER

CREATE PROXIMITY ZONE AROUND THE RIVER (BUFFER TOOL)

CREATE PROXIMITY ZONES AROUND EXISTING PARKS (BUFFER TOOL)

SUBTRACT PARK PROXIMITY ZONES FROM RIVER PROXIMITY ZONE (OVERLAY: ERASE TOOL)

ASSIGN DEMOGRAPHIC VALUES FROM BLOCK GROUPS TO RIVER PROXIMITY ZONE (OVERLAY: IDENTITY TOOL)

FAVORABLE

NOT FAVORABLE

SELECT AREAS WITH FAVORABLE DEMOGRAPHIC VALUES (ATTRIBUTE QUERY)

SELECTED

NOT SELECTED

SELECT VACANT PARCELS IN AREAS WITH FAVORABLE DEMOGRAPHICS (SPATIAL QUERY)

• COUNT
• NO COUNT

BUFFER SELECTED PARCELS AND COUNT CENSUS BLOCK POPULATIONS (SPATIAL JOIN)

RECOMMENDED SITES

3000 PEOPLE LIVE NEARBY

ASSIGN POPULATION COUNTS TO PARCELS (OVERLAY: IDENTITY TOOL)

Looking at the sketch of the analysis brings up the point that this plan represents only one of several possible approaches to the problem. The plan we're following here is, we think, a reasonable and efficient one. It uses a variety of important geoprocessing tools and has the visual benefit of letting you see the analysis unfold, as land areas are progressively stripped away from consideration. It is certainly possible, however, to reach the same or similar conclusions with different combinations of tools.

Your analysis results depend on the state of the data in the ReadyData geodatabase. If you've completed all the exercises up to now, you should be in good shape. If necessary, download the Lesson 5 results from the Understanding GIS web site. These results include the ReadyData geodatabase and its contents in the correct state to continue the project.

Exercise 6a: Establish proximity zones

In this exercise, we have two objectives. First, we want to define the area of interest as a ¾-mile zone around the river. Any parcels falling outside this zone will be excluded from consideration. Second, we want to draw ¼-mile zones around each park. Any parcels in the area of interest that lie within these internal exclusion zones will also be dropped. We'll use one simple operation—buffer—to significantly reduce our hunting grounds.

1) **Create a results geodatabase.** When we generate outputs we need to store them somewhere. In Lesson 4, we decided to keep our input data in one geodatabase and our output data in another geodatabase. (See the topic *Project database considerations.*) It's a matter of preference, but we think it's easier to keep the project organized if inputs and outputs are separated. We may want to share the input data with someone else so they can run the analysis independently. We may want to repeat the analysis with different parameters (something that will happen in Lesson 7). The more feature classes you add to a geodatabase, the harder it is to keep track of what they represent and what purpose they serve.

Ⓐ Start ArcMap and open a blank map.

Ⓑ In the Catalog window, expand Folder Connections and navigate to the folder C:\UGIS\ParkSite\AnalysisData.

Ⓒ Right-click the AnalysisData folder and choose New → File Geodatabase.

Ⓓ Rename the geodatabase **AnalysisOutputs** (Figure 6-2).

▷ How? The name should be editable when you create the geodatabase. If not, you can right-click it and choose Rename.

Figure 6-2   Name the new geodatabase AnalysisOutputs

We need to make AnalysisOutputs the default geodatabase so our geoprocessing outputs go there automatically.

E In the Catalog window, right-click AnalysisOutputs and choose Make Default Geodatabase.

2) **Buffer the LA River.** A buffer is a zone around a map feature measured in distance units. We'll use the Buffer tool to create a ¾-mile proximity zone around the LA River. For more information about the Buffer tool and other analysis tools, see the topic *Essential GIS analysis tools*.

A In the Catalog window, under AnalysisData, expand ReadyData.gdb and add *LARiver* to the map.

Note that we added the *LARiver* layer to an empty map display without putting a contextual basemap underneath it. Also, the layer is likely symbolized in a color other than blue. During analysis, we're more interested in geoprocessing results than we are in map appearance. For that reason, we're not going to symbolize input and output layers carefully at each step along the way. We'll do it as the need arises, but often we'll just accept the ArcMap defaults.

B From the main menu, choose Geoprocessing ➔ Buffer.

C In the Buffer dialog box, click the Input Features drop-down arrow and choose *LARiver*.

D Rename the output `LARiverBuffer`.

E In the Linear Unit box, type `0.75`. Click the drop-down arrow for units and choose Miles.

F Double-check your settings (Figure 6-3) and click OK.

Figure 6-3

When the tool finishes, the *LARiverBuffer* layer is added to the map.

Naming conventions revisited

In our project database considerations, we established naming conventions for the ReadyData geodatabase. We'll adhere to these conventions in AnalysisOutputs with respect to feature classes that we intend to keep and use in our final map. However, most of the geoprocessing outputs in this lesson will be "intermediate" data: essential steps in the process, but not ends in themselves. Eventually, intermediate data will be deleted from the geodatabase. With respect to naming these datasets, we'll follow an expedient approach. If the default ArcMap name is short and not confusing, we'll use it. If we need to clarify a dataset's purpose or analytical meaning in the course of operations, we'll assign it a more suitable name.

Essential GIS analysis tools

Analysis tools

The tools shown here are by no means a complete list, but they include several of the ones used most often in GIS analysis. It's hard (and probably unnecessary) to give an exact definition of what makes a tool an "analysis" tool. GIS practitioners solve problems of many different kinds, and most of these problems have facets that involve spatial relationships among geographic objects: how far is A from B, how many As are close to B, which areas are common to A and B, how do you get from A to B, and so on. Analysis tools quantify these relationships among features and their attributes.

Query tools

Query tools answer questions of the form "Which features meet such-and-such a condition?" Select By Attributes selects features according to an attribute value or combination of values. Select By Location selects features according to their spatial relationship to features in another layer (or sometimes the same layer). Spatial relationships include intersection, containment, adjacency, and distance.

Query tools can be accessed from the Selection menu in ArcMap or as geoprocessing tools (Select Layer By Attribute and Select Layer By Location).

Proximity tools

The Buffer tool creates a feature class of polygons at a specified distance around input features. It is often used to draw an exclusion zone around features (no A's should be allowed within a mile of B), or to define an area of interest (we want to look for A's only within a mile of B).

The Create Thiessen Polygons tool creates a feature class of contiguous polygons around input point features. Each polygon's shape is defined by proximity to the nearest point. Thiessen polygons can be used to define allocation areas (which areas are closer to A than to any other point, which are closer to B than to any other point, and so on).

Select By Attributes

Which cities have a population of 100,000 or more?

Select By Location

Which parks are within a mile of the river?

Create Thiessen Polygons

Buffer

Streams have quarter-mile buffers.

Input points are parks within a river buffer. Each output polygon encloses the area that is closer to the point it contains than to any other point.

Essential GIS analysis tools (continued)

Proximity tools (continued)

The Near tool finds the nearest feature in one or more specified layers to each feature in the input layer. It writes the distance as a new attribute to the input layer table.

Near

Near

The input features are vacant parcels (symbolized in red). The features to find are parks (green). New attributes in the vacant parcels table specify the nearest park to each vacant parcel and its distance.

OBJECTID *	Shape *	NEAR_FID	NEAR_DIST
1	Polygon	117	1371.816526
2	Polygon	114	822.280138
3	Polygon	101	1234.614597
4	Polygon	34	2465.358086
5	Polygon	35	2012.800925

A table relate shows the attributes of the nearest park to the selected parcel.

OBJECTID *	Shape *	NAME	TYPE
117	Polygon	Greaver Oak Park	Local park or recreation area

Overlay tools

The basic purpose of overlay tools is to find common areas between different layers. Overlays answer questions about where A and B, each with unique and important attributes, overlap. In contrast to queries, which return selections of existing features, overlay tools create new features that have the attributes of both input layers.

The different overlay tools—Intersect, Union, Identity, and Erase are the most common—perform variations on the same basic process. They differ with respect to how much area from the input layers is included in the output feature class: common area only (Intersect), all area whether common or not (Union), one input layer's area only (Identity), or one input layer's area minus the common area (Erase).

Parks and stream buffers have areas of overlap.

Intersect

Spatial Join

Spatial Join is like a Select By Location query with the added benefit of joining source layer attributes to target layer features.

The overlay creates new features (purple) from areas of overlap. The output can be used to find park land as a percentage of riparian land or vice versa.

Spatial Join

Vacant parcels (red) within a quarter mile of the river are spatially joined to the river. In the output feature class, the river has the same geometry but new attributes: 400 vacant parcels totaling 1,232 acres satisfy the quarter-mile condition.

OBJECTID *	Shape *	Join_Count	TARGET_FID	NAME	ACRES	Shape_Length
1	Polyline	400	1	Los Angeles River	1232.55	166495.37503

G Confirm the presence of the new feature class by expanding the Catalog window and looking in AnalysisOutputs.gdb (Figure 6-4).

Figure 6-4

The new feature class you created

3) **Buffer the parks to ¼ mile.** Now we'll add the *Parks* data and draw a quarter-mile buffer around each park. These buffers will encompass areas that are close to existing parks and therefore out of consideration.

A In the Catalog window, under ReadyData, drag *Parks* to the map window.

B In the Table of Contents, right-click *LARiverBuffer* and choose Zoom To Layer.

C Turn off the *LARiver* layer.

Figure 6-5

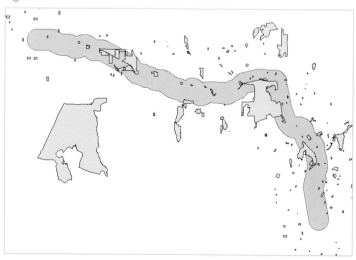

We can already see a lot of overlap between existing parks and the river buffer zone (Figure 6-5). We still need to extend the exclusion zone a quarter-mile further out on all sides of each park.

D From the main menu, choose Geoprocessing→ Buffer.

E Set the Input Features to *Parks*.

F Accept the default Output Feature Class name of *Parks_Buffer*.

G In the Linear Unit box, type **0.25**. Set the units to Miles.

H Click the Dissolve Type drop-down arrow and choose ALL.

This setting dissolves boundaries between overlapping park buffers. The dissolution of boundaries isn't strictly required, but it makes it a lot easier to interpret the result, which would otherwise be a dense tangle of lines. A side effect of dissolving the buffers is that the output will consist of one multipart feature, essentially devoid of attributes. That's okay—all we need from this layer is its geometry.

Figure 6-6

Set to ALL to dissolve boundaries.

❶ Check your settings against Figure 6-6 and click OK.

When the tool finishes, a *Parks_Buffer* layer is added to the map and a feature class is created in AnalysisOutputs.

❷ Turn off the *Parks* layer.

❸ Open the layer properties for *Parks_Buffer* and click the Display tab.

❹ In the Transparent % box, type **50**. Click OK.

The portion of *LARiverBuffer* not covered by the *Parks_Buffer* layer represents the areas that are still under consideration. (The layers are purple and green in Figure 6-7, but your colors may be different.)

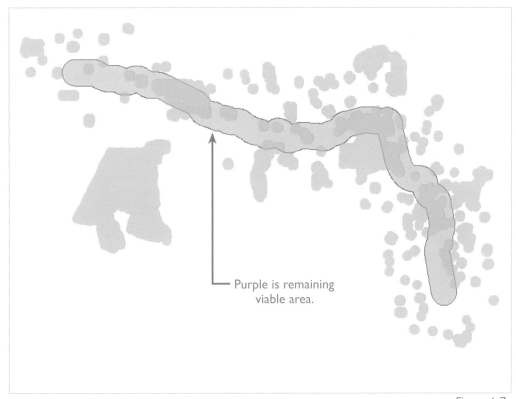

Purple is remaining viable area.

Figure 6-7

Let's think about what we have. One layer, *LARiverBuffer,* contains all the area within ¾ mile of the river. We want to exclude the *Parks_Buffer* areas from within that, but those areas are stored in a different layer. Essentially, we want to subtract one layer from another.

4)   Erase park accessible areas. The tool for doing this subtraction is named Erase. This tool is a member of the family of spatial overlay operations that help make GIS such a powerful technology. Unlike the Buffer tool, the Erase tool is not available from the Geoprocessing menu, so we'll find it with a search.

Ⓐ Open the Search window.

  ▷ If the Search tab isn't present, on the Standard toolbar, click the Search button 🔍. Set the window to Auto Hide.

Ⓑ Near the top of the Search window, click Tools to restrict your search.

Ⓒ In the Search box, start to type **erase**. When "erase (analysis)" appears in the drop-down list (Figure 6-8), click on it to search for it.

Ⓓ In the list of results, click Erase (Analysis), as shown in Figure 6-9, to open the tool.

Ⓔ For the Input Features, choose *LARiverBuffer.*

Ⓕ For the Erase Features, choose *Parks_Buffer.*

Ⓖ Name the Output Feature Class **ProximityZone**.

Ⓗ Check your settings against Figure 6-10 and click OK.

Figure 6-8

Figure 6-9

Figure 6-10

When the process is finished, the *ProximityZone* layer is added to the map.

❶ In the Table of Contents, turn off all layers except *ProximityZone*.

❿ Symbolize *ProximityZone* in a medium blue (for example, Cretan Blue). This will help it stand out against the basemap we'll add in the next step.

We're left with a reduced portion of the original buffer zone (Figure 6-11). Our search for a park is now confined to this shape.

Erase subtracts the green areas from the purple.

ProximityZone is the geometric result of the Erase operation.

Figure 6-11

5) Add a basemap. This new feature is quite an abstract shape, and we can't forget that it represents real territory on the ground. We'll add a basemap to get some geographic orientation.

Ⓐ Click the Add Data drop-down arrow ⊞▾ and choose Add Basemap.

Ⓑ In the list of basemaps, add the Topographic basemap.

C Open the layer properties for *ProximityZone*. Click the Display tab, if necessary.

D Change the transparency to **50**% and click OK.

E Zoom in on some of the "holes" in the *ProximityZone* layer.

You can see the effect that the existing parks (visible on the basemap) had on the creation of the feature (Figure 6-12).

Figure 6-12

F Pan along the river and look at a few more examples.

If you see some anomalies—such as a hole with no park—consider that the topographic basemap may not correspond exactly to the *Parks* layer. You can turn the *Parks* layer on and off in the Table of Contents for reference.

G When you're finished, remove the *Basemap* layer and zoom to the *ProximityZone* layer.

H Save the map document as **Lesson6** in the MapsAndMore folder.

I If you are continuing to the next exercise now, leave ArcMap open. Otherwise, exit ArcMap.

In the next exercise we'll evaluate the neighborhood demographics within the area of interest.

Exercise 6b: Apply demographic constraints

We've isolated areas that meet two of our requirements: distance to the river and distance from parks. Still ahead are the tasks of factoring in neighborhood demographics (population density, income, and age) and evaluating parcels by size and total population served. We'll deal with the demographics in this exercise, and with the parcels and population served in Exercise 6c.

Within the proximity zone, we want to find areas that have the right demographic criteria. The problem is that our blue blob, if we can call it that, doesn't include those attributes; they are found only in the *BlockGroups* feature class. What we want to do, then, is make a feature class that has the spatial area of *ProximityZone* and the attribute values of *BlockGroups*.

This again calls for an overlay. In the last exercise, we subtracted one layer from another, but overlay operations are typically more additive than subtractive. Usually, we want to find common ground between layers that have different attributes. The important attribute in *ProximityZone* is its distance to the river. The important attributes in *BlockGroups* are population density, age, and median income. By overlaying these two layers, we'll get all these important attributes in a single layer.

The overlay will cause any block group features that lie partly inside and partly outside the proximity zone to be split. What happens to the attribute values of these features? If a block group with 2,000 people is halfway in the proximity zone, should the output feature defined by the overlapping half get all 2,000 of those people, or just 1,000? By default, attribute values are copied rather than apportioned in overlay operations, and this can lead to anomalies where large attribute values are assigned to small slivers of geometry. We've dealt with this problem ahead of time by using block group attribute values that are statistically homogenized and assumed to be uniform across the feature. It's also possible, however, to redistribute attribute values by area during geoprocessing. See the topic *Apportioning attribute values.*

1) **Overlay block groups on the proximity zone.** We want to keep all the geometry of the *ProximityZone* layer and only as much *BlockGroups* geometry as is spatially coincident with the proximity zone. This type of overlay is called Identity: it keeps all the geometry of Layer A and just the coincident geometry of Layer B.

   Ⓐ If necessary, start ArcMap and open Lesson6 from the list of recent maps.

   Ⓑ In the Catalog window, add *BlockGroups* from ReadyData.gdb.

   Ⓒ In the Search window, highlight and delete the existing search term. Start to type **identity**.

D Choose "identity (analysis)" from the drop-down list to locate it.

E In the list of results, click Identity (Analysis) to open the tool.

F For the Input Features, choose *ProximityZone*.

G For the Identity Features, choose *BlockGroups*.

H Accept the default Output Feature Class name of *ProximityZone_Identity*.

I Check your settings against Figure 6-13 and click OK.

Figure 6-13

When the tool finishes, the *ProximityZone_Identity* layer is added to the map. A corresponding feature class is created in the AnalysisOutputs geodatabase.

J Zoom to the *ProximityZone_Identity* layer and turn off all other layers in the Table of Contents.

What we see in Figure 6-14 is a fractured version of the original proximity zone. *ProximityZone_Identity* has the geometry of *BlockGroups* clipped to the boundaries of *ProximityZone*. As we'll see, it has the attributes of both input layers.

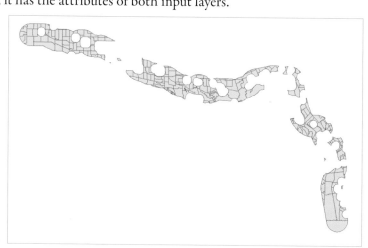

Figure 6-14

2) **Examine attributes.** Looking at the attributes of both our input and output layers will help us understand what we've accomplished.

Ⓐ Open the *ProximityZone* layer attribute table (Figure 6-15).

Figure 6-15

The BUFF_DIST value is expressed in feet

It consists of a single record. Its NAME and BUFF_DIST (buffer distance) attributes are inherited from the *LARiverBuffer* layer.

Ⓑ Keep the table window open. Open the attribute table for the *BlockGroups* layer (Figure 6-16). Scroll across its attributes.

This layer has 6,421 records. It has the attributes we specified and created in Lesson 4.

Figure 6-16

Ⓒ Keep the table window open. Open the attribute table for the *ProximityZone_Identity* layer.

Ⓓ At the bottom of the table window, drag and drop the ProximityZone_Identity tab onto the blue arrow pointing upward (Figure 6-17).

These values match OBJECTID values in *BlockGroups*.

These values match OBJECTID values in *ProximityZone*.

Figure 6-17

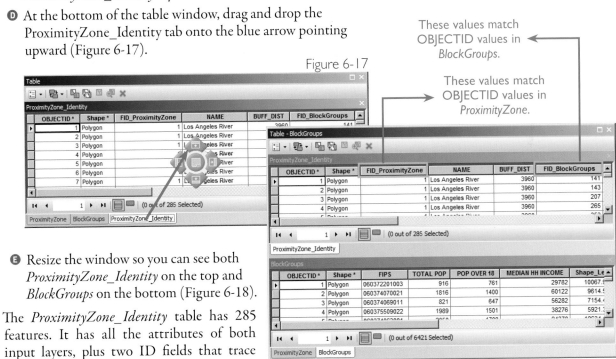

Ⓔ Resize the window so you can see both *ProximityZone_Identity* on the top and *BlockGroups* on the bottom (Figure 6-18).

The *ProximityZone_Identity* table has 285 features. It has all the attributes of both input layers, plus two ID fields that trace the features back to the input features with which they are spatially coincident.

Figure 6-18

F Keep the table window open but move it away from the map.

G Zoom in on the western end of the *ProximityZone_Identity* layer.

 ▷ A map scale of around 1:24,000 is good.

H Turn on the *BlockGroups* layer.

I On the Tools toolbar, click the Select Features tool 🔍.

J Select the *ProximityZone_Identity* feature indicated in Figure 6-19, or any other feature that is cut by the proximity zone boundary.

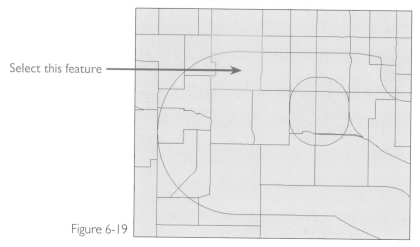

Select this feature ——————

Figure 6-19

You can see on the map and in the tables that two features are selected: one in *ProximityZone_Identity* and one in the *BlockGroups* layer underneath.

K In the table window, click the Show Selected Records button 🔲 for both tables.

L Scroll the two tables so that some common attributes line up (Figure 6-20).

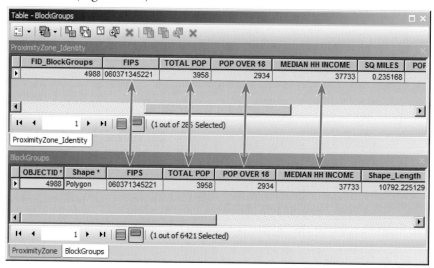

Figure 6-20

All the block group values have been copied to the spatially corresponding features in the *ProximityZone_Identity* layer. That's what we want—these cut-up block groups (some are cut up, some are intact) are the "neighborhoods" that we'll query for suitable demographics.

Note the importance of analyzing attributes such as density and percentage, rather than raw numbers. The POP UNDER 18 value of the selected block group is 1,024. We would reasonably expect this value to be smaller in the *ProximityZone_Identity* feature (because it doesn't include the whole block group), but because the attribute values are copied, it's the same. It's fair to assume, however, that the PCT UNDER 18 value would remain the same even when the block group is cut into smaller pieces.

ⓜ On the Tools toolbar, click Clear Selected Features 🔲.

ⓝ Close the table window.

ⓞ Turn off the *BlockGroups* layer.

ⓟ Zoom to the *ProximityZone_Identity* layer.

3) Select areas by demographic attributes. The *ProximityZone_Identity* layer now contains the demographic attributes of interest: population density, the percentage of children, and median household income. We want to consider areas only if they meet our thresholds for all three values:

- Population density greater than or equal to 8,000 people per square mile
- Percentage of children greater than or equal to 23%
- Median household income less than or equal to $50,000

We'll use an attribute query to select features that satisfy these conditions.

ⓐ From the main menu, choose Selection → Select By Attributes.

ⓑ At the top of the Select By Attributes dialog box, click the Layer drop-down menu and choose *ProximityZone_Identity*.

ⓒ Make sure the Method is set to "Create a new selection."

ⓓ Build the following expression:

```
"POPDENSITY" >= 8000 AND "PCTUNDER18" >= 23
         AND "MEDHINC" <= 50000
```

Combining the queries in a single statement is more efficient than making a series of selections and subselections. The AND operator ensures that features will be selected only by passing *all* the tests.

ⓔ Check your expression against Figure 6-21 and click OK.

Figure 6-21

The map shows the features that satisfy the query. There aren't that many and most of them lie toward the ends of the proximity zone (Figure 6-22).

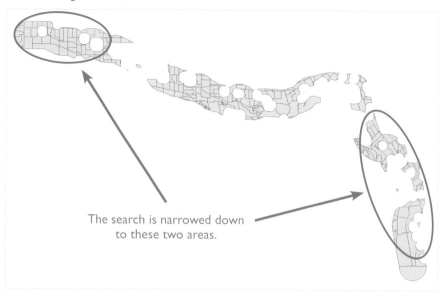

The search is narrowed down to these two areas.

Figure 6-22

F Open the *ProximityZone_Identity* attribute table.

G Confirm that there are 25 selected records and close the table.

4) Create a feature class from the selected features. We'll save these selected features to a new feature class. Previously, we used the Copy Features tool for this purpose, but this time we'll use the Export Data command. The advantage of this command is its convenient location on the layer context menu. Export Data isn't incorporated into the geoprocessing framework, which means it doesn't respect environment settings, but that's not a concern in this case.

Figure 6-23

A In the Table of Contents, right-click *ProximityZone_Identity* and choose Data ➜ Export Data.

B Highlight and delete the default output feature class name (Export_Output). Replace it with **GoodZones**.

▷ Deleting the path and restoring it with the Tab key won't work in this dialog box.

C Check your settings against Figure 6-23 and click OK.

D Click Yes on the prompt to add the data as a layer.

E On the Tools toolbar, click the Clear Selected Features button ⊠ and turn off the *ProximityZone_Identity* layer.

F Save the map.

G If you are continuing to the next exercise now, leave ArcMap open. Otherwise, exit ArcMap.

Exercise 6c: Select suitable parcels

We've narrowed our focus down to a layer called *GoodZones*. Having covered the demographic criteria, we'll turn our attention to the parcels. Thanks to our work in Lesson 4, we have a dataset consisting of all and only vacant parcels. From this dataset, we'll select the parcels that lie within good zones. From that selection, we'll subselect parcels that are one acre or larger: that will comprise our set of candidate park sites. Then we'll go on to do some further analysis to compare the sites by total population served and distance to the river.

1) **Select vacant parcels within good zones.** We'll select the parcels with a spatial query.

   Ⓐ If necessary, start ArcMap and open Lesson6 from the list of recent maps.

   Ⓑ In the Catalog window, add *VacantParcels* from ReadyData.gdb.

   Ⓒ From the main menu, choose Selection → Select by Location.

   Ⓓ At the top of the Select By Location dialog box, make sure the Selection Method is set to "select features from."

   Ⓔ Under Target Layer(s), check the box next to *VacantParcels*.

   Ⓕ Click the Source Layer drop-down arrow and choose *GoodZones*.

   Ⓖ Confirm that the spatial selection method for target layer feature(s) is set to "intersect the source layer feature."

   Ⓗ Confirm that the box to apply a search distance is unchecked.

Parcels will be selected if they lie partially or entirely within good zones.

Some parcels may cross good zone boundaries because of how the zones are clipped. Should we include parcels that straddle good zones? There's no right or wrong answer. By choosing the "intersect" spatial selection method, we're going with a generous interpretation—we'll select a parcel if any part of it lies within a good zone because we want to maximize the candidate pool. If we wanted to be restrictive (the parcel has to be *entirely* within the good zone), we'd choose a method like "[target layer feature(s)] are within the source layer feature." See Appendix A for further discussion of this issue.

   Ⓘ Check your settings against Figure 6-24 and click OK.

Figure 6-24

Figure 6-25

① Open the *VacantParcels* attribute table.

You should have just 88 of 15,802 parcels selected.

2) Subselect parcels one acre or larger. Of the 88 selected parcels, we want to keep only those that are at least one acre. We'll do this with an attribute query on the ACRES field.

Ⓐ At the top of the table, click the Table Options button 〔≡〕 and choose Select By Attributes.

Ⓑ In the Select By Attributes dialog box, click the Method drop-down arrow and choose "Select from current selection."

This restricts the scope of the query to records that are already selected.

Ⓒ Build the following expression: "ACRES" >= 1.

Ⓓ Check your dialog box against Figure 6-25. Click Apply and close the Select By Attributes dialog box.

We're down to seven selected records: just seven parcels meet all our conditions.

Ⓔ In the Table of Contents, right-click *VacantParcels* and choose Selection→Zoom To Selected Features.

Ⓕ In the attribute table, click the Show selected records button 〔≣〕 (Figure 6-26).

Ⓖ Close the attribute table.

At this scale, it's hard to see the selected parcels on the map. We'll take a closer look in the next step.

Figure 6-26

3) **Export selected features.** These seven parcels are the critical ones we want: they represent the culmination of the process, and we want to save them as a feature class. Again, we'll use the convenient Export Data command.

Ⓐ In the Table of Contents, right-click *VacantParcels* and choose Data→ Export Data.

Ⓑ Highlight and delete the default output feature class name (Export_Output). Replace it with **SevenSites**.

Ⓒ Check your settings against Figure 6-27 and click OK.

Ⓓ Click Yes to add the exported data as a layer.

4) **Inspect the candidates.** Now we'll zoom in and take a close look at the seven sites against basemap imagery.

Ⓐ Add the *Imagery* basemap layer.

Ⓑ In the Table of Contents, remove *VacantParcels.* Turn off all layers except *SevenSites* and *Basemap.*

Ⓒ In the Table of Contents, open the Symbol Selector for the *SevenSites* layer.

Ⓓ Set the fill color to No Color and the outline color to Medium Apple (Figure 6-28). Make the outline width 2. Click OK.

Ⓔ Open the attribute table for the *SevenSites* layer and move it away from the map.

Ⓕ In the table, right-click the gray box next to the first record and choose Zoom To (Figure 6-29).

The map zooms to the first candidate parcel (Figure 6-30).

Figure 6-27

Figure 6-28

Figure 6-29

Right-click and choose Zoom To on the context menu.

Figure 6-30

G On the Standard toolbar, click the Scale Box drop-down arrow and choose 1:10,000.

H If necessary, pan to the southwest so you can see the river.

Notice that three of our seven sites are right next to each other. (We'll come back to this point in the next step.) Notice also that these sites aren't especially close to the river.

I In the Table of Contents, turn on the *ProximityZone* layer and the *Parks* layer.

J Optionally, symbolize the *Parks* layer in a shade of green (Figure 6-31).

Figure 6-31

The view confirms that the sites are near the outer edge of the proximity zone. Even more interesting is that another park is closer to the river on a direct line. That's why human interpretation of analysis results is so important. We considered proximity to existing parks, but it didn't occur to us that a park might be more than a quarter mile from a candidate site and yet closer to the river. It doesn't invalidate our candidates, but it's something to think about.

K Turn off the *ProximityZone* and *Parks* layers.

L In the table, right-click the gray box next to the fourth record and choose Zoom To.

M Zoom to a map scale of **1:2,500** (Figure 6-32).

Figure 6-32

This site is right next to the river, which is ideal. Notice a couple of other things. First, the river bottom is natural here, which could make this site more scenic than others. Our analysis doesn't take aesthetic considerations into account, but the views, sounds, and smells of a park affect the people who use it.

On the other hand, the site is also adjacent to a freeway. This has an aesthetic impact and probably a health impact, as well: should kids be exposed to all that car and truck exhaust? The freeway is also a barrier to accessibility. We haven't yet done the part of the analysis that considers the total population served, but our formula uses a simple distance buffer that doesn't consider freeways or other barriers (see Appendix A).

This site is next to railroad tracks.

N Zoom to the remaining sites (Figure 6-33) and investigate them:
- Zoom and pan to see the site in relation to the river.
- Turn other map layers on and off.
- Make observations that add context to the analysis.

O When you're finished, turn off all layers except *SevenSites* and *Basemap*.
  ▷ Leave the attribute table open.

5) Aggregate sites. Our first three sites are separated by narrow gaps. These are probably unpaved easements, or rights-of-way, intended to become streets in the event of development. Because the sites are so close together, and so much alike, it seems odd to present them as distinct choices. We can't dissolve them—they don't share boundaries—but we can achieve a similar effect with a tool called Aggregate Polygons. This tool merges nearby polygon features on the basis of a specified distance value. We'll make measurements to get a sense of what this value should be, and then aggregate the sites to give the city council a "large park" option.

This site has an irregular shape.

A In the attribute table, zoom to the first record.

B If any records are selected, clear them with the Clear Selection tool ⌷ at the top of the table.

C Close the attribute table.

D Zoom out a little and pan until the three nearby parcels more or less fill your screen.

E On the Tools toolbar, click the Measure tool ▦.

F On the Measure dialog box, make sure the Measure Line tool is selected (Figure 6-34).

G Click the Choose Units drop-down arrow and choose Distance ➔ Feet.

This site is partially outside the proximity zone.

Figure 6-34

Figure 6-33

H Draw lines to measure the distances at a few different places along the gaps between the parcels (Figure 6-35).

Measure the gaps between these parcels

Figure 6-35

You should get measurements of about 30 feet.

I Close the Measure dialog box.

Before we go ahead with the aggregation, let's acknowledge the assumptions we've made about land use: that any vacant parcel is available regardless of ownership, that it's okay to combine adjacent parcels, and now that it's okay to combine nearly adjacent parcels. Any of these assumptions may turn out to be problematic in reality, but it seems reasonable to suppose that the acquisition of land for a public park might require buying multiple parcels.

Unlike the other tools we're using in this lesson, Aggregate Polygons is found under Cartography Tools in the Generalization toolset. The main purpose of generalization tools is to simplify or refine the appearance of features for small-scale map display—our use of the tool in this context is somewhat unorthodox.

J Open the Search window. If necessary, click Tools near the top. Delete the existing search term, if any and start to type **aggregate**.

K On the drop-down list of tools, choose "aggregate polygons (cartography)" to locate it.

L In the list of results, click Aggregate Polygons (Cartography) to open the tool.

▷ Be careful not to choose Aggregate Points.

M In the Aggregate Polygons dialog box, set the Input Features to *SevenSites*.

N Name the Output Feature Class **FiveSites**.

O Enter an Aggregation Distance of **50** and make sure the units are Feet.

We're allowing a little margin of error.

P Check your settings against Figure 6-36 and click OK.

When the tool finishes, a layer called *FiveSites* and a table called *FiveSites_Tbl* are added to the Table of Contents. On the map, you can see the new, aggregated polygon on top of the three original sites (Figure 6-37).

The Table of Contents has switched to the List By Source view to show the table.

Q In the Table of Contents, open *FiveSites_Tbl*.

Figure 6-36

This backlink table shows which of the features in *SevenSites* were aggregated. We might need it under other circumstances, but not here.

R Close *FiveSites_Tbl* and remove it from the Table of Contents.

S At the top of the Table of Contents, click the List by Drawing Order button 📇.

T Open the *FiveSites* attribute table.

We're now down to five candidates. (Only the three sites in the view were aggregated because none of the others were near each other.) Note that we've lost the ACRES attribute from the *SevenSites* layer. A consequence of aggregating features, as with dissolving them, is the loss of attributes.

6) Calculate acreage. To restore acreage values, we'll add a field and calculate its geometry.

Figure 6-37

A In the *FiveSites* attribute table, click the Table Options button 📇 and choose Add Field.

B Name the field **ACRES** and set the Type to Double (Figure 6-38). Click OK.

C In the attribute table, right-click the ACRES field and choose Calculate Geometry.

  ▷ You won't get a warning if you turned warnings off in Lesson 4. If you do, click Yes to continue.

D In the Calculate Geometry dialog box, set the units to Acres US [ac] and accept the other defaults. Click OK.

E If necessary, click Yes on the warning about making calculations outside an edit session.

Figure 6-38

The field is populated with acreage values (Figure 6-39).

Figure 6-39

F Close the Attribute table.

At this point, there are two items remaining from our list of analysis criteria: we want the new park to serve the maximum number of people and we want it to be as close to the river as possible.

7) **Define park access zones.** In Lesson 2, we defined park accessibility in terms of a quarter-mile distance. This is an oversimplification of accessibility, but it has the virtue of being easy to analyze: we just have to buffer the candidate sites and then count the population inside them.

A In the Table of Contents, turn off the *SevenSites* layer.

B Right-click the *FiveSites* layer and choose Zoom to Layer.

You probably can't see the sites anymore, but that's okay.

C From the main menu, choose Geoprocessing→Buffer.

D In the Buffer dialog box, set the Input Features to *FiveSites*.

E Name the Output Feature Class **ParkAccessZones**.

F In the Linear Unit box, type **0.25** and set the units to Miles.

G Double-check your settings against Figure 6-40 and click OK.

Figure 6-40

When the tool finishes, the *ParkAccessZones* layer is added to the map. You can see the five roughly circular features representing the quarter-mile access zones around the candidates (Figure 6-41).

Figure 6-41

8) Add block centroids to the map. We need to find the population within each of these access zones. The basic approach of counting the block centroids in each zone and summing their populations was introduced in Lesson 2 and sketched in the Analysis Plan at the beginning of this lesson. For another way, see the topic *Apportioning attribute values.*

How do we do it? With a spatial query, we could select block centroids within park access zones. Then, in the *BlockCentroids* table, we could right-click the POP2010 field, choose Statistics, and see the sum of the selected records. It would mean doing five operations (each access zone has to be queried separately), but that's not a big deal. A bigger drawback is that this method returns the answer in a message box—we want to write it to a table.

We can do that using the Spatial Join tool. You might think of a spatial join as a spatial query plus a table join: attributes from one table are joined to another on the basis of a spatial relationship between layers, rather than a common attribute.

Let's get a visual sense of what we're going to accomplish.

Figure 6-42

Ⓐ Turn off the *Basemap* layer.

Ⓑ Add *BlockCentroids* to the map from ReadyData.gdb.

Ⓒ Drag *ParkAccessZones* to the top of the Table of Contents.

Ⓓ Symbolize *ParkAccessZones* with a fill color of No Color. Make the outline width 2 and choose any outline color you want.

Ⓔ Zoom in on one of the access zones (Figure 6-42).

Ⓕ Open the *BlockCentroids* attribute table (Figure 6-43).

The POP2010 field values will be summed for block centroids within each access zone and written to a new feature class attribute table.

Ⓖ Close the *BlockCentroids* table.

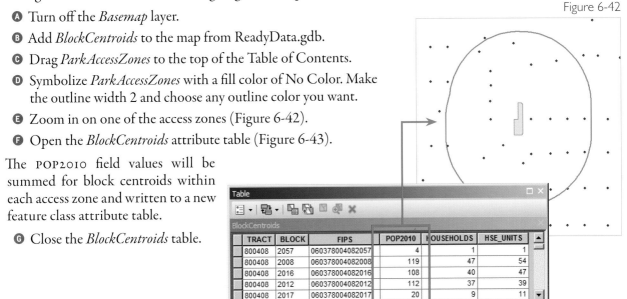

Figure 6-43

	TRACT	BLOCK	FIPS	POP2010	HOUSEHOLDS	HSE_UNITS	
	800408	2057	060378004082057	4	1	1	
	800408	2008	060378004082008	119	47	54	
	800408	2016	060378004082016	108	40	47	
	800408	2012	060378004082012	112	37	39	
	800408	2017	060378004082017	20	9	11	

Figure 6-44

The only *BlockCentroids* attribute we need is POP2010. We'll turn the other fields off so they're not carried forward.

ⓗ Open the layer properties for *BlockCentroids* and click the Fields tab.

ⓘ Click the Turn all fields off button ⬚.

ⓙ In the list of fields, check the POP2010 box to turn it back on (Figure 6-44) and click OK.

We don't need any of the attributes from the *ParkAccessZones* layer.

ⓚ Open the layer properties for *ParkAccessZones*. On the Fields tab, click the Turn all fields off button (Figure 6-45). Click OK.

9) Spatially join block centroids to park access zones. We're ready to do the spatial join. Unlike a table join, a spatial join isn't virtual—the output is a new feature class.

ⓐ Open the Search window. Highlight and delete the existing search term and begin typing **spatial join**.

ⓑ On the drop-down list of tools, choose "spatial join (analysis)" to locate it.

ⓒ In the list of results, click Spatial Join (Analysis) to open the tool.

ⓓ In the Spatial Join dialog box, set the Target Features to *ParkAccessZones*.

ⓔ Set the Join Features to *BlockCentroids*.

ⓕ Accept the default Output Feature Class name of *ParkAccessZones_SpatialJoin*.

In the Field Map area, the only fields are POP2010 and the required length and area fields.

ⓖ In the Field Map area, right-click POP2010 and choose Merge Rule → Sum (Figure 6-46).

Figure 6-45

Figure 6-46

Click Sum

The merge rule setting is crucial. It tells ArcMap that for each target feature (access zone) we want to sum the population values of the join features (block centroids).

ⓗ Below the Field Map area, click the Match Option drop-down arrow and choose CONTAINS.

▷ You may need to resize the dialog box or scroll down to see all the parameters.

ⓘ Compare your dialog box to Figure 6-47 and click OK.

When the tool finishes, a *ParkAccessZones_SpatialJoin* layer is added to the map.

ⓙ Open the attribute table for the *ParkAccessZones_SpatialJoin* layer (Figure 6-48).

The POP2010 field has the total population within a quarter mile of each site. We've written these values to a table, which is what we wanted, but we haven't written them to the *FiveSites* table, which is really *where* we want them. We'll take care of this in just a little bit.

ⓚ Close the table.

10) Calculate distance to the river. Our last requirement is to figure out the distance from each site to the LA River. We could use the Measure tool to do that, of course, but we'd face the same problem of writing the results to a table. Also, we might not be sure of making the shortest possible measurement. The Near tool will make these calculations for us automatically.

ⓐ Open the Search window. Delete the existing search term and type **near**.

ⓑ On the drop-down list of tools, choose "near (analysis)" to locate it.

Figure 6-47

Figure 6-48

The number of block centroids in each access zone

The population within 0.25 miles of each site

The park access zone ID

⊙ In the list of results, click Near (Analysis) to open the tool.

⊙ In the Near dialog box, set the Input Features to *FiveSites*.

⊙ Set the Near Features to *LARiver*.

Note that you can add multiple layers to the list of Near features. Typically, the tool is used to discover which feature, in one or more layers, is closest to a feature of interest. In our case, we just want the distance measurement.

The Near tool doesn't create a new feature class. Instead, it will add a couple of fields to the *FiveSites* table.

⊙ Check your settings against Figure 6-49 and click OK.

⊙ When the tool finishes, open the *FiveSites* layer attribute table and scroll to the right (Figure 6-50).

Figure 6-49

Figure 6-50

The NEAR_DIST field stores the distance, in feet, of each site to the river.

⊙ Close the table.

11) Add the park-accessible population to *FiveSites*. It would be nice to store all the relevant site-selection attributes in a single table with the candidate sites:

- Acreage
- Distance to the river
- Total population within a quarter mile
- Demographic variables

That will make it easy to evaluate and compare the sites at a glance—for us and for anyone we share the data with. We already have the acreage and distance attributes in the *FiveSites* table. In this step, we'll add the population totals that we got from the spatial join in Step 9.

We could do this with another spatial join, but we'll go back to the Identity tool as a matter of preference. As we've done before, we'll filter out the attributes we don't need.

Ⓐ Open the layer properties for *ParkAccessZones_SpatialJoin*.

Ⓑ On the Fields tab, click the Turn all fields off button ⬚.

Ⓒ Check the POP2010 box to turn it back on (Figure 6-51) and click OK.

Ⓓ Open the Identity (analysis) tool.

Ⓔ For the Input Features, choose *FiveSites*.

Ⓕ For the Identity Features, choose *ParkAccessZones_SpatialJoin*.

Ⓖ Accept the default Output Feature Class name of *FiveSites_Identity*.

Ⓗ Check your settings against Figure 6-52 and click OK.

When the tool finishes, the *FiveSites_Identity* layer is added to the map.

Ⓘ Open the *FiveSites_Identity* layer attribute table and scroll to the right (Figure 6-53).

Figure 6-51

Figure 6-52

Figure 6-53

The POP2010 values (representing population within a quarter mile) have been brought in.

Ⓙ Close the table.

Choose which fields will be visible

☐ OBJECTID
☐ Shape
☐ FIPS
☐ TOTAL POP
☐ POP OVER 18
☑ MEDIAN HH INCOME
☐ Shape_Length
☐ Shape_Area
☐ SQ MILES
☑ POP DENSITY
☐ POP UNDER 18
☑ PCT UNDER 18

Figure 6-54

12) Add the demographic attributes to *FiveSites_Identity*. We'll use Identity once more to get the demographic attributes into the table. These attributes are found in the *BlockGroups* layer.

Ⓐ Open the layer properties for *BlockGroups*.

Ⓑ On the Fields tab, click the Turn all fields off button.

Ⓒ Check the following fields to turn them back on:

- MEDIAN HH INCOME
- POP DENSITY
- PCT UNDER 18

Check your settings against Figure 6-54 and click OK.

Ⓓ Open the Identity (analysis) tool.

Ⓔ For the Input Features, choose *FiveSites_Identity*.

Ⓕ For the Identity Features, choose *BlockGroups*.

Ⓖ Rename the Output Feature Class **RecommendedSites**.

Ⓗ Check your settings against Figure 6-55 and click OK.

Figure 6-55

When the tool finishes, the *RecommendedSites* layer is added to the map.

Ⓘ Open the *RecommendedSites* layer attribute table and scroll across it.

It has all the attributes of interest, plus several ID attributes (backlinks to features in other tables) that we don't need. In the next exercise, we'll format this table to make it presentable.

Ⓙ Close the table.

Ⓚ Save the map.

Ⓛ If you are continuing to the next exercise now, leave ArcMap open. Otherwise, exit ArcMap.

Apportioning attribute values

Splitting feature geometry has implications for attribute values. In Example 1, a census block group is split by a buffer in an overlay operation, resulting in two output features. By default, ArcMap copies the attribute values of the input feature to both output features. That's okay for POPDENSITY, which is already a ratio: the value applies to the parts as well as to the whole. It's not okay for TOTALPOP, a count value, because it doubles the population, as shown.

In Example 2, we want to count the population inside the buffer, which covers parts of three block groups. A reasonable approach is to analyze the percentage of each block group's area that falls inside the buffer and assign the same percentage to its population value. In this case, 1% of block group 1's area is inside the buffer, so its contribution to the buffer population should be 1% of 1,755. Likewise for block group 2 on the far right (21% of 4,445) and block group 3 in the middle (38% of 6,722).

The examples are variations of the same problem, which can be solved with a tool named Make Feature Layer. This tool makes a layer from a feature class (essentially the same thing that happens automatically when you add data to ArcMap). It has a very useful parameter, however: a "ratio policy" check box for each attribute in the input feature class. When you geoprocess a layer made with this tool, the ratio policy is applied to any split features, dividing their attribute values according to area.

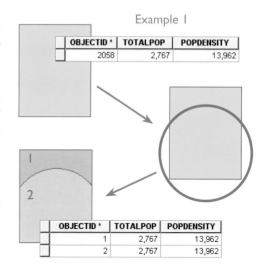

Example 1

OBJECTID *	TOTALPOP	POPDENSITY
2058	2,767	13,962

OBJECTID *	TOTALPOP	POPDENSITY
1	2,767	13,962
2	2,767	13,962

Example 2

OBJECTID	TOTALPOP
1	1,755
2	4,445
3	6,722

The "Use Ratio Policy" box is checked for the TOTALPOP field.

Make Feature Layer is run on the *BlockGroups* feature class. Subsequent overlay operations give desired results.

Example 1 solution

OBJECTID *	TOTALPOP	POPDENSITY
1	762	13,962
2	2,005	13,962

Example 2 solution

OBJECTID *	TOTALPOP
1	16
2	934
3	2,535

The final step to get the total population of the buffer area would be to use the Dissolve tool with TOTALPOP as a statistics field.

Exercise 6d: Clean up the map and geodatabase

Analysis projects tend to leave clutter behind. It's normal to find your working map and geodatabase filled with a mixture of data you want to preserve and data you can now discard. It's tempting to leave the mess behind, but sooner or later someone will return to the scene and wish that you'd taken the time to put things in order. In this exercise, we'll simplify our map document, which contains several unnecessary layers. We'll also format the *FiveSites* layer attribute table and save it as a layer file. Finally, we'll remove intermediate data from the geodatabase and add some metadata to the datasets we want to keep.

1) **Clean up the map document.** This isn't the map document we'll use for our final layout, but we still want it to be intelligible. With that in mind, we'll keep just a few important layers in it.

Ⓐ If necessary, start ArcMap and open Lesson6.mxd.

Ⓑ In the Table of Contents, remove all layers except the four listed below. Put the layers in the top-to-bottom order shown:

- *RecommendedSites*
- *LARiver*
- *LARiverBuffer*
- *Basemap*

Ⓒ Turn all four layers on and zoom to the *LARiverBuffer* layer.

Ⓓ Symbolize *RecommendedSites* with a fill color of No Color, an outline width of 2, and an outline color of Medium Apple (Figure 6-56).

Ⓔ Symbolize the *LARiver* layer with the color Big Sky Blue and a width of 3.

Ⓕ Symbolize *LARiverBuffer* with a fill color of No Color, an outline width of 3, and an outline color of Medium Lilac.

Your map should look like Figure 6-57.

Figure 6-56

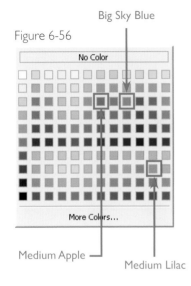

Big Sky Blue

Medium Apple

Medium Lilac

Figure 6-57

2) **Add layer descriptions.** When you add a layer to a map, a description—taken from the item description of the source data—is added to the layer properties. This description explains what the layer represents. Layer descriptions are helpful to anyone using the map document and are required if you share layers as packages. The *LARiver* layer already has a description, as does the *World Imagery* sublayer of the basemap; the other two, created during geoprocessing, do not.

A Open the layer properties for *RecommendedSites* and click the General tab.

B In the Description box, type `These are the five sites recommended for a new park near the Los Angeles River.` (Figure 6-58) Click OK.

C Open the layer properties for *LARiverBuffer*. In the Description box, type `This is a 3/4 mile buffer around the Los Angeles River.` (Figure 6-59) Click OK.

Figure 6-58

3) **Delete unnecessary fields.** In the last exercise, we got all the important attributes into the *RecommendedSites* table. We also have several FID attributes that we don't need to maintain.

A Open the *RecommendedSites* attribute table.

B In the table, right-click the FID_FIVESITES_IDENTITY field heading (Figure 6-60) and choose Delete Field.

C Click Yes on the prompt to confirm the field deletion.

The field is deleted from the table.

Figure 6-59

Delete this field

Be careful when you delete a field because you can't undo it.

Figure 6-60

D In the same way, delete all the remaining FID fields from the table. They are:

- FID_FIVESITES
- NEAR_FID
- FID_PARKACCESSZONES_SPATIALJOIN
- FID_BLOCKGROUPS

The table should look like Figure 6-61.

Figure 6-61

OBJECTID *	Shape *	ACRES	NEAR_DIST	POP2010	MEDIAN HH INCOME	POP DENSITY	PCT UNDER 18	Shape_Length	Shape_Area
1	Polygon	5.683245	3086.15668	3427	42291	16546.38	34.67012	2603.230091	247561.150336
2	Polygon	1.658465	191.625914	1475	39049	8109.447	32.08092	1329.288752	72242.454828
3	Polygon	1.950284	2537.569999	3107	34893	8044.595	30.32706	1702.573374	84954.047345
4	Polygon	3.433686	2482.656555	1338	39166	8185.089	25.13417	1767.268413	149570.782296
5	Polygon	1.711671	3603.387356	3820	46464	13948.06	28.99952	1493.750791	74560.107748

4) **Add a field.** We're going to add an identifier field that we can manage. We'll have a reason to use this field in Lesson 8.

A Click the Table Options button 🔲 and choose Add Field.

B Name the field **SiteID** and accept the default Short Integer field data type.

C Check your settings against Figure 6-62 and click OK.

D Scroll to the end of the table. Right-click the SITEID field heading and choose Field Calculator.

▷ If necessary, click Yes on the warning.

E In the Field Calculator dialog box, double-click OBJECTID in the list of fields to add it to the expression box.

Figure 6-62

This expression will assign the same values of 1 to 5 to the new field. The difference is that SITEID is a field we can manage, while OBJECTID is not.

F Check your settings against Figure 6-63 and click OK.

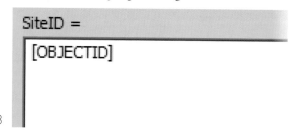

Figure 6-63

The field is populated with the values.

G Leave the table window open, but move it out of the way.

5)   Format fields. Most of the fields in the table should be given an alias or formatted in other ways to make them more readable. We'll go through the steps in detail for one field and let you do the rest with general instructions.

Ⓐ Open the layer properties for *RecommendedSites* and click the Fields tab.

Ⓑ In the list of fields, uncheck SHAPE_LENGTH and SHAPE_AREA to hide them.

Ⓒ At the bottom of the list, click SITEID to highlight it.

Ⓓ In the row of tools at the top, click the little drop-down arrow next to the Move Up button ↑ ▾ and choose Move To Top.

Ⓔ Click the Move Down button ↓ ▾ twice to move the field directly underneath SHAPE (Figure 6-64).

Moving this field affects the layer only. In the feature class, fields are stored in the order in which they're added, and this order can't be changed.

Arrange the display order of fields with these buttons

Turn these fields off

Figure 6-64

Ⓕ Click the PCT UNDER 18 field near the bottom of the list to highlight it.

Ⓖ In the box on the right, under Appearance, click the word "Alias" to highlight it. Replace the current alias with **Percent Under 18 Years**.

Ⓗ Two lines down, click Number Format to highlight it.

The next three steps are shown as a sequence in Figure 6-65.

Figure 6-65

Click to format as %

Change to zero decimal places

Ⓘ Click the button that appears on the right to open the Number Format dialog box.

Ⓙ In the Number Format dialog box, click the Percentage category and click Numeric Options.

Ⓚ Change the number of decimal places to 0.

Ⓛ Click OK on the Numeric Options and Number Format dialog boxes.

Ⓜ Click Apply to apply your changes so far.

N Format the remaining fields according to the following table. Depending on the field, you may have to

- Change or add an alias.
- Change the number format.
- Change the number of decimal places.
- Check the box to show thousands separators.

Field	Alias	Number Format
ACRES	Acres	Numeric (one decimal place)
NEAR_DIST	Feet to LA River	Numeric (zero decimal places; show thousands separators)
POP2010	1/4 Mile Population	Numeric (show thousands separators)
MEDIAN HH INCOME	Median Household Income	Currency
POP DENSITY	People per Square Mile	Numeric (zero decimal places; show thousands separators)

O Click OK to close the Layer Properties.

P Widen fields as needed by placing the mouse pointer between field headings and dragging in either direction.

Your table should look like Figure 6-66.

Figure 6-66

Q If necessary, open the layer properties and make any needed changes.

R Close the attribute table.

6) **Save the layer as a layer file.** All these format settings are layer properties. If we want them to be available in other map documents, we need to save *RecommendedSites* as a layer file, as we did with the city boundary of Los Angeles in Lesson 1.

A In the Table of Contents, right-click *RecommendedSites* and choose Save As Layer File.

B In the Save Layer dialog box, navigate if necessary to Home - ParkSite\MapsAndMore.

Ⓒ Accept the default name *RecommendedSites.lyr* (Figure 6-67) and click Save.

7) **Set map document properties.** Maps, like datasets, can have metadata. By filling out a few fields, you can supply some background information about your map. This is helpful to anyone you share the map document with and is required if you decide at some point to publish the map as an online web service.

Figure 6-67

Ⓐ From the main menu, choose File → Map Document Properties.

Ⓑ In the Map Document Properties dialog box, for the Title, type `Los Angeles River Park Site Analysis`.

Ⓒ For the Summary, type `Map of recommended sites for park development near the Los Angeles River.`

Ⓓ For the Description, type `These are the five sites recommended for a new park near the Los Angeles River.`

Ⓔ For the Author, type your name.

Ⓕ For the Credits, type `City of Los Angeles, Esri.`

Ⓖ For the Tags, type `parks, Los Angeles River, Los Angeles.`

▷ Include the commas to separate the tags.

Ⓗ Near the bottom of the dialog box, check the box to store relative paths to data sources.

Figure 6-68

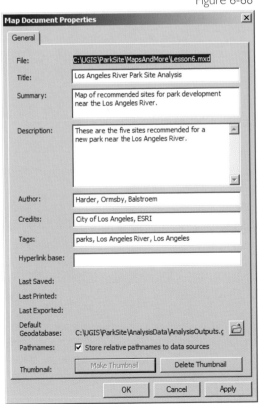

This setting allows you to move the map document and its data sources to different locations without disruption, as long as the layer paths to source data maintain the same directory structure relationship.

Ⓘ Click Make Thumbnail.

Ⓙ Compare your settings to Figure 6-68 and click OK.

Ⓚ Save the map.

Ⓛ In the Catalog window, in the Home folder, right-click *Lesson6.mxd* and choose Item Description.

The properties you set appear in the Item Description for the map document. (If you don't see the thumbnail graphic, right-click the Home folder in the Catalog window and choose Refresh.) Note that the Preview tab allows you to navigate the map.

Ⓜ Close the Item Description window.

8) **Clean up the geodatabase.** In this step, we'll delete intermediate data from the geodatabase and update the item descriptions for the two feature classes that we're keeping.

Ⓐ In the Catalog window, expand AnalysisOutputs.gdb and click on it.

Ⓑ If necessary, click the Toggle Contents Panel button 🔳 to split the Catalog window into a top and bottom pane.

The geodatabase should contain eleven feature classes and one table, as shown in Figure 6-69.

Ⓒ In the bottom pane of the Catalog window, hold the Control key and click each dataset in AnalysisOutputs to select it, except these two:

- *LARiverBuffer*
- *RecommendedSites*

Ⓓ Compare your Catalog window to Figure 6-70 and right-click any of the selected datasets and choose Delete.

Figure 6-69

Figure 6-70

Ⓔ Click Yes to delete the datasets.

Now we'll update the Item Descriptions for the two remaining feature classes. The ones they have now have been copied from their geoprocessing input datasets.

Ⓕ In the Catalog window, right-click the *LARiverBuffer* feature class and choose Item Description.

G In the Item Description window, click the Preview tab and click the Create Thumbnail button 🖫.

H Click the Description tab. In the row of buttons at the top, click Edit.

I In the Summary box, highlight and delete the existing text. Replace it with this: **This dataset represents the area under consideration for a new park near the Los Angeles River.**

J In the Description box, highlight and delete the existing text. Replace it with this: **This dataset is a 3/4 mile buffer around the Los Angeles River.**

K In the Credits box, highlight and delete the existing text. Replace it with this: **City of Los Angeles, Esri.**

L At the top of the window, click Save (Figure 6-71).

M Leave the Item Description window open. In the Catalog window, click *RecommendedSites*.

The Item Description window updates.

N Click the Preview tab.

At this scale the sites are hardly visible, which wouldn't make a very good thumbnail image,

O On the Item Description toolbar, click the Zoom In tool ⊕ and zoom in on one of the sites (Figure 6-72).

P Click the Create Thumbnail button 🖫.

Q Click the Description tab and click Edit.

R In the Title box, replace the existing text with this: **Recommended Sites**.

S In the Summary box, replace the existing text with this: **These are the five sites recommended for a new park near the Los Angeles River.**

T In the Description box, replace the existing text with this: **This dataset is a subset of vacant land parcels one acre or larger near the Los Angeles River.**

U In the Credits box, replace the existing text with this: **City of Los Angeles, Esri.**

V Click Save and close the Item Description window.

W Save the map.

Figure 6-71

Figure 6-72

Exercise 6e: Evaluate results

The recommended sites meet our criteria and look good on basemap imagery, but ultimately, there's no substitute for on-site inspection. How do these candidate locations really look in the context of their surroundings? If you had the opportunity, it would be worthwhile to visit the sites and record your impressions. In this exercise, we've included some photographs of the sites to help give a sense of that experience.

1) Compare your guesses from Lesson 1. In Lesson 1, you made some guesses about likely areas for parks based on a preliminary look at the data. Let's see how those guesses turned out.

- Ⓐ In the Catalog window, under the Home folder, drag and drop *Lesson1Predictions.tif* onto the map.
  - ▷ Close the Geographic Coordinate Systems Warning if you get it.
  - ▷ If you don't have this file, you can download and extract it from the Lesson 1 results at the Understanding GIS web site.
- Ⓑ Zoom to the *Lesson1Predictions.tif* layer (Figure 6-73).

Figure 6-73

Lesson1Predictions.tif is a raster dataset. Its GeoTIFF tags contain the coordinate system information ArcMap needs to align it with the other layers in the map.

How did your guesses turn out? You can zoom and pan as needed to get a better look. (The raster layer will lose resolution as you zoom in.)

Ⓒ When you're finished, turn off the *Lesson1Predictions.tif* layer.

Converting graphics to features

Another way to make the comparison is to convert the black ellipses from graphic elements to features. Here are the steps:

- Open Lesson1.mxd.
- On the main menu, click Edit and choose Select All Elements.
- Add the Draw toolbar.
- On the Draw toolbar, click the Drawing drop-down menu and choose Convert Graphics to Features.
- Save the feature class in AnalysisOutputs.gdb.
- Add the new feature class to the Lesson6 map.

2) Examine candidates. We'll take a short virtual tour of the five candidate sites.

Ⓐ Open the *RecommendedSites* attribute table, and zoom to Site 1 (Figure 6-74), then zoom to Site 2 (Figure 6-75).

Figure 6-74

This site has good demographics but is far from the river.

The site is hilly with limited road access.

Figure 6-75

This site is adjacent to the river. It has a high percentage of children but relatively low population density and total population.

The site is also right next to a freeway.

B Zoom to Site 3 (Figure 6-76) and zoom to Site 4 (Figure 6-77).

Figure 6-76

This site has the lowest median household income, but is otherwise more or less in the middle of the demographic pack. It also has a developed residential parcel jutting into it.

The site is apparently used for storage and is next to high-voltage power lines.

Figure 6-77

This site is perhaps the weakest overall in terms of demographics.

The site is now being used informally as a soccer field.

⊙ Zoom to Site 5 (Figure 6-78).

Figure 6-78

This site has a high population density and the highest quarter-mile population, but is the farthest site from the river, actually crossing the 3/4-mile buffer.

The site is easily accessed from streets.

⊙ Close the attribute table and zoom to the *LARiverBuffer* layer.

⊙ Save the map and exit ArcMap.

This wraps up our analytical work. We started with a set of (fictional) guidelines from the city council. Those guidelines were vague in many cases, so we ended up supplying our own working definitions. Obviously, these could be questioned. The guidelines also excluded many factors that might be important: whether the land is actually available, its environmental condition, barriers to access such as freeways, and several more. A number of caveats to the analysis are discussed in Appendix A. Finally, what the council did not give us—and we have no way to evaluate on our own—is the relative importance of the criteria. We can't rank our candidates because we don't know, for example, if low household income outweighs a high percentage of children in making a site desirable, and if so, by how much. We have to present the options neutrally because we don't know what the council's priorities are. (They may not know either until they begin to study the results.)

Suppose somebody asked you how the results would change if you set the income requirement lower or the percentage of children requirement higher? How could you answer a question like that without rerunning the entire analysis from scratch? Read on to Lesson 7 for the solution.

Lesson

7 Automate the analysis

SUPPOSE UNEXPECTED CHANGES occur after you've finished the analysis. Maybe an updated table of vacant parcels becomes available. Maybe after seeing the results, the City Council wants to add or remove criteria, or see how a change in threshold values affects the outcome. Of course, you could redo the analysis, but even a very small change—especially if it comes at the beginning of the process—could make for a lot of tedious work. This is the moment to introduce ModelBuilder™. ModelBuilder can encapsulate your entire workflow in a single geoprocessing tool that runs with the push of a button.

A model looks like a flowchart or schematic diagram that shows the steps for manufacturing a product. Inputs, outputs, and processes are represented by geometrical shapes. Arrows connect the shapes to show the sequence of operations. For an example, see the topic *Understanding a model*.

A model is more than just a visual aid; it's also an engine. The schematic elements of the model embed the datasets and tools they represent, so that running a model sets in motion a chain of geoprocessing operations. In our case, the model will basically repeat the analysis tasks we did in Lesson 6.

There are lots of good reasons to build a model from our analysis. First, being a diagram, a model makes the whole workflow transparent and open to scrutiny. This may reveal its weaknesses or confirm its validity—either way, it brings the methodology into the open. Second, a model stores, and thereby documents, the workflow, sparing you the task of keeping detailed notes on what you did and why you did it. Third, models are flexible. They easily accommodate the kinds of changes to inputs or parameters mentioned above with little effort on your part. Fourth, models can be shared with colleagues, saving development work and encouraging collaboration on sound methods and best practices. Fifth, models run pretty fast—just wait and see.

Lesson Seven roadmap

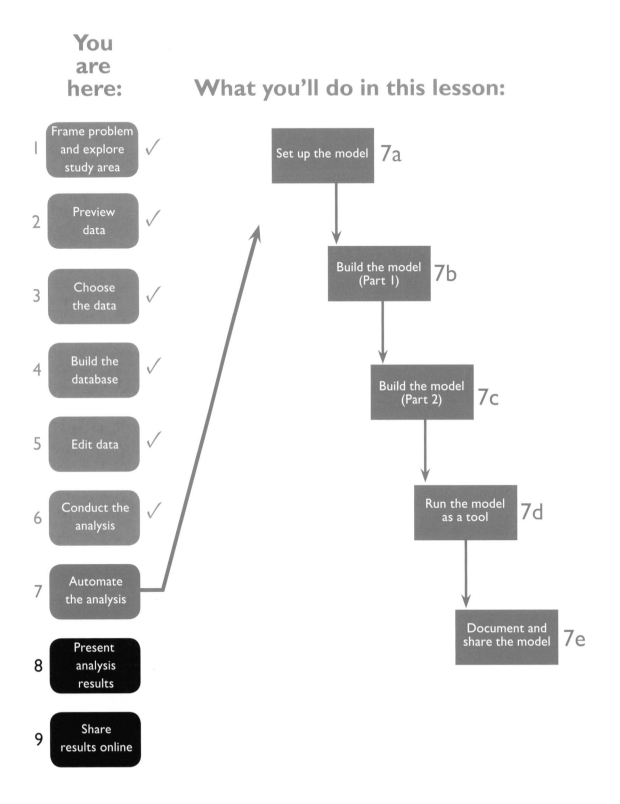

You are here:

1 Frame problem and explore study area ✓

2 Preview data ✓

3 Choose the data ✓

4 Build the database ✓

5 Edit data ✓

6 Conduct the analysis ✓

7 Automate the analysis

8 Present analysis results

9 Share results online

What you'll do in this lesson:

Set up the model — 7a

Build the model (Part 1) — 7b

Build the model (Part 2) — 7c

Run the model as a tool — 7d

Document and share the model — 7e

Exercise 7a: Set up the model

ModelBuilder can be opened as an application from within either ArcMap or ArcCatalog. We'll use ArcMap for the same reasons we ran geoprocessing tools from ArcMap in Lesson 6: it's easy to add map layers as tool inputs and it's useful to be able to look at outputs in the map.

1)   Create an output geodatabase. As in Lesson 6, we'll create a new geodatabase to hold our output data.

Ⓐ Start ArcMap and open a new blank document.

Ⓑ Open the Catalog window and pin it down to keep it open.

▷ If the Catalog tab isn't showing, click the Catalog window button 🔳 on the Standard toolbar.

Ⓒ Expand Folder Connections and navigate to C:\UGIS\ParkSite.

Ⓓ Right-click the AnalysisData folder and choose New → File Geodatabase.

Ⓔ Rename the new geodatabase **ModelBuilderOutputs**.

Ⓕ Right-click the ModelBuilderOutputs geodatabase and choose Make Default Geodatabase (Figure 7-1).

Figure 7-1

Ⓖ Save the map document (never mind that it's empty) in the MapsAndMore folder and name it **Lesson7**.

Understanding a model

To get the basic idea, let's forget about the LA River for a moment and prepare a Saturday morning breakfast. At right is a recipe (workflow) for making a dozen classic American pancakes from scratch. You should pick up the ingredients at your local store and make a batch to celebrate when you have completed this workbook!

The ingredients are symbolized by blue ellipses (inputs). These are prepared in various ways as shown by the orange rectangles (tools or processes). The results are green ellipses (outputs). You stir the basic ingredients to make the flour mix before adding the whipped egg, oil, and buttermilk. The mix should be properly blended before liquids are added; otherwise, you may get clumps of baking powder in the dough, which will lead to some very flat pancakes. After you make the dough, refrigerate it for several hours (or, even better, overnight) to allow time for chemical reaction. Then fry the dough in a pan with your favorite fat and eat it with a good brand of maple syrup. Delicious!

Can anything go wrong? If you follow the arrows, you're on a tried and true path. If you start improvising, you might make the recipe even better—but you might also get your weekend off to an indigestible start.

It's much the same with an ArcGIS workflow. With a complex analysis, it usually takes trial and error to get the recipe right. Typically, you'll experiment with several combinations of data, tools, and sequences before finding the best way to solve your problem. Once you find it, you want to save it and share it with friends, just like the pancake recipe.

We experimented a lot with the data and many ArcGIS tools before writing this cookbook (sorry, workbook). Having figured out the workflow, we implemented it in Lesson 6. But we also wanted to preserve it, encapsulate it, make it a formula—as you'll see, that's what ModelBuilder is all about.

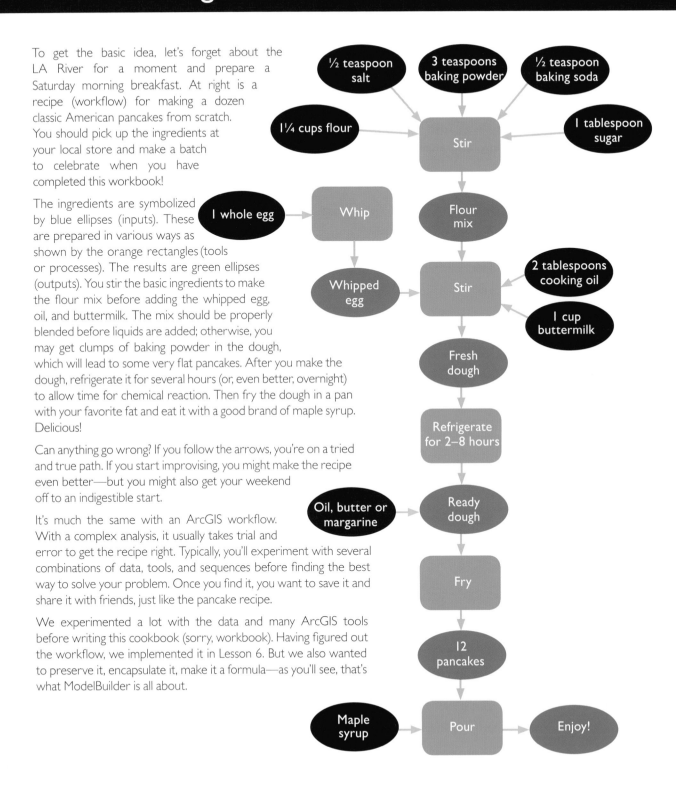

2) Create a toolbox and model. A model is a type of tool, and, as such, needs to go in a toolbox. It can't go in a system toolbox, however. We'll create a new toolbox in the MapsAndMore folder and put it there. We could also put it in My Toolboxes (under Toolboxes, at the root level of the Catalog window) or in any other folder or geodatabase.

Ⓐ At the top of the Catalog window, right-click the Home folder (ParkSite\MapsAndMore) and choose New → Toolbox.

Ⓑ Rename the new toolbox **ParkPlanningModels**.

Ⓒ Right-click the ParkPlanningModels toolbox and choose New → Model.

An empty Model window opens (Figure 7-2). This window can be resized and moved outside the ArcMap application window.

Figure 7-2

Ⓓ In the Catalog window, expand ParkPlanningModels.tbx, if necessary.

The model exists as a tool in the toolbox (Figure 7-3).

3) Set model properties. We'll rename the model and supply some information about it.

Ⓐ In the Model window, from the main menu, choose Model → Model Properties.

Ⓑ In the Model Properties dialog box, click the General tab.

Ⓒ Replace the default name with **ParkSuitabilityAnalysis** (no spaces).

Figure 7-3

D Replace the default label with **Park Suitability Analysis** (this time include spaces).

The label is like a field name alias: a name that's more descriptive or easier to read.

E For the Description, enter **This model identifies vacant parcels near the Los Angeles River that may be suitable for a new park.**

F Optionally, add other information that will help identify the model, such as your name, your organization, and the date.

G Check the box to store relative pathnames.

The model, its input data, and its output data can now be moved without disruption as long as they remain in the same workspaces relative to each other.

H Make sure the "Always run in foreground" box is checked.

Ordinarily, geoprocessing tools run in the background so that you can keep using ArcMap while they work. Running the model in the foreground means that you have to stop and wait while the model executes. That's normally advisable because models are computationally intensive.

I Check that your entries match Figure 7-4 and click OK.

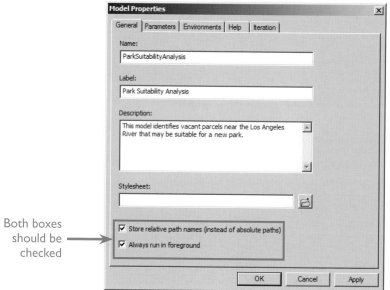

Both boxes should be checked

Figure 7-4

J On the ModelBuilder toolbar, click the Save button 🖫 to save the model.

In the Catalog window, the model now appears with its label.

K Optionally, unpin the Catalog window in ArcMap.

Exercise 7b: Build the model (Part I)

Our model will re-create the analysis workflow from Lesson 6. Knowing how much work that entailed should give you insight into why models are such an efficient tool: not only for analysis projects, but for any series of operations that you might want to save, repeat, vary, or use as a basis for further development. Re-creating the analysis will let you focus your attention on the modeling process itself, since the data and geoprocessing tools will be familiar.

The data preparation tasks from Lesson 4 won't be part of the model. A big advantage of models is that it's easy to rerun them. We don't want to incorporate operations such as adding fields, calculating field values, joining tables, and so on, that only need to be done once.

We're also going to leave out the step in Lesson 6 in which three nearby parcels were aggregated. That step was based on a human judgment that the gaps between the parcels were easements (unbuilt roads) and that it was legitimate to consolidate the parcels. We can't build that sort of judgment into the model.

In this exercise, we'll model about half of the analysis steps, saving the remainder for the next exercise. The stopping point is more or less arbitrary, but building a model incrementally is a good practice. It's easier to find and fix problems if you do test runs along the way.

1)   **Add data to ModelBuilder.** In this step, we'll add all the input data for the model. This is the opposite of what we did in Lesson 6, where we added layers one at a time as we needed them. Either way is fine in either ArcMap or ModelBuilder: it's a matter of preference.

- Ⓐ On the ModelBuilder toolbar, click the Add Data or Tool button ✛.
  - ▷ If you closed the model: in the Catalog window, right-click *Park Suitability Analysis* and choose Edit.
  - ▷ If you closed ArcMap: start ArcMap, open the document Lesson7.mxd, then edit the model as described above.
- Ⓑ In the Add Data or Tool dialog box, navigate to C:\UGIS\ ParkSite\AnalysisData\ReadyData.gdb.
- Ⓒ Hold down the Control key and click these five feature classes (Figure 7-5):
  - *BlockCentroids*
  - *BlockGroups*
  - *LARiver*
  - *Parks*
  - *VacantParcels*
- Ⓓ Click Add.

Figure 7-5

Figure 7-6

Figure 7-7

Each dataset is added as a blue ellipse to the Model window. (These blue ellipses are "input data variables," but we'll mostly just call them inputs.) The elements aren't automatically sized to fit; you may only be able to see a couple of them.

E Enlarge the Model window to about three-fourths of your screen space.

F On the ModelBuilder toolbar, click the Full Extent button ⬚.

You should see all five elements centered in the Model window and marked with blue selection handles.

The display of text within ellipses changes as you zoom in or out in the Model window. The places where words divide may change and so may the number of lines of text. Don't worry if the text display on your screen is different from the figures.

G Hover over any selected element; then click and drag the whole group to the left side of the Model window.

H Click in some empty white space in the Model window to unselect the elements (Figure 7-6).

2) **Add a tool to ModelBuilder.** The first step of the analysis in Lesson 6 was to buffer the LA River. Let's start with that.

A On the ModelBuilder toolbar, click the Add Data or Tool button ⊕.

B Click the Look in drop-down arrow and choose Toolboxes (Figure 7-7).

C Double-click System Toolboxes, Analysis Tools, then the Proximity toolset.

D Click the Buffer tool to highlight it (Figure 7-8) and click Add.

Figure 7-8

The tool is added to the Model window, connected by a small arrow to an Output Feature Class (Figure 7-9). The tool and its output are white because they're not ready to run. The tool needs to be connected to input data and its parameters have to be filled out.

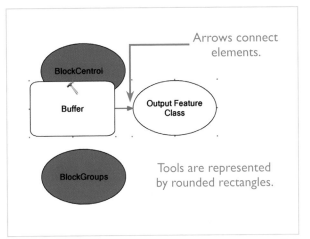

Figure 7-9

ⓔ Make sure the Buffer tool and Output Feature Class elements are selected (marked with blue selection handles).

▷ If they're not, hold down the Shift key and click to select them. You don't have to select the arrow.

ⓕ Drag the selected elements down and place them to the right of the *LARiver* input data element. (The exact position doesn't matter.)

ⓖ On the ModelBuilder toolbar, click the Connect tool 💠 and move the mouse pointer over the blue *LARiver* ellipse.

The cursor looks like a magic wand ⚟ .

ⓗ From the *LARiver* ellipse, drag and drop a blue arrow on top of the Buffer tool (Figure 7-10) and release the mouse button.

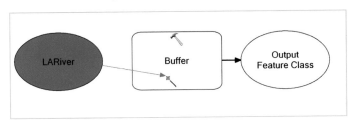

Figure 7-10

You are prompted for the type of connection to make (Figure 7-11).

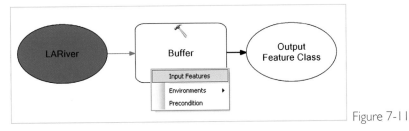

Figure 7-11

ⓘ Click Input Features.

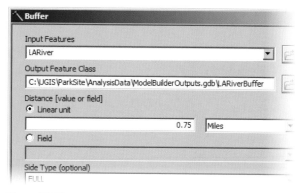

Figure 7-12

J Right-click the Buffer tool and choose Open.

The familiar geoprocessing tool opens. The input features are set to *LARiver,* as specified by the connection arrow, and the output feature class is directed to the default geodatabase. We'll rename this feature class in accordance with our convention.

K Highlight and delete the existing output feature class path. Type **LARiverBuffer** and press Tab.

   ▷ Or just remove the underscore from "LARiver_Buffer."

L For the linear unit, type **0.75**. Set the adjacent drop-down arrow to Miles (Figure 7-12). Click OK.

In ordinary geoprocessing, clicking OK would run the process; in ModelBuilder it *prepares* the process to run. The tool and output data elements turn yellow and green, respectively, which means they are properly connected and filled out (Figure 7-13).

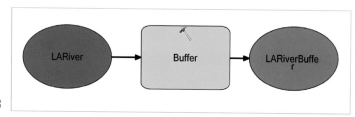

Figure 7-13

We now have a process ready to run—a very simple model—but we're not going to run it yet.

The green element displays the name of the output feature class. In the figure above, the line breaks badly, stranding the letter "r" on a line by itself. This effect is a function of many things: the zoom level, the text length, the font, and the size of the ellipse. It probably looks the same on your screen, but even if it doesn't, you'll see the effect elsewhere. We'll fix it by changing the text. (We could also resize the ellipse or just leave things as they are—it's purely a display issue in the Model window.)

M Right-click the *LARiverBuffer* element. On the context menu, choose Rename.

N In the Rename box, replace the existing name with **LA River Buffer** (Figure 7-14). Click OK.

Figure 7-14

The text is reformatted. Word spaces help the line break naturally (Figure 7-15).

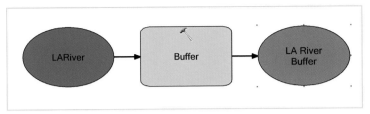

Figure 7-15

This name will be given to layers added to ArcMap from the model, but it won't change the output feature class name.

○ Right-click the green *LA River Buffer* output and choose Open (Figure 7-16).

Figure 7-16

display name

output feature class

The display name has spaces but the output feature class name doesn't.

○ Close the LA River Buffer dialog box.

○ On the ModelBuilder toolbar, click the Save button 💾 .

3) **Buffer the parks.** In Lesson 6, after buffering the river, we created ¼-mile buffers around existing parks.

○ On the ModelBuilder toolbar, click the Add Data or Tool button ✛.

The Add Data or Tool dialog box should open to the Proximity toolset, and the Buffer tool should be right there.

○ Click the Buffer tool to select it and click Add.

As before, the tool and its output feature class are added to the model. The tool is named Buffer (2) to distinguish it from the other Buffer tool.

○ On the ModelBuilder toolbar, click the Select tool ↖ .

○ Drag the selected elements down and place them to the right of the *Parks* input data element.

  ▷ If you don't click the Select tool, and your cursor is still a magic wand, you'll draw a connector. Press the Escape key to delete it.

Figure 7-17

Last time, we set the input features in the Buffer tool by drawing a connector. Now, we'll do it another way.

E Right-click the Buffer (2) tool and choose Open.

▷ Or double-click the tool to open it.

F Click the Input Features drop-down arrow and choose *Parks*.

The available inputs are marked with a blue recycle symbol (for "data variables") in contrast to the familiar symbol for layers—there actually aren't any layers in the map document. Every input data element is technically a variable because you can open it and change the feature class it references.

G For the linear unit, type **0.25**. Set the units drop-down arrow to Miles.

H Click the Dissolve Type drop-down arrow and choose ALL.

We don't have the visual reason to dissolve the buffers that we had in Lesson 6 (we're not going to look at this tool's output), but it doesn't cause us any trouble to duplicate the workflow.

I Compare your dialog box to Figure 7-17 and click OK.

Both buffer processes should now be colored (Figure 7-18).

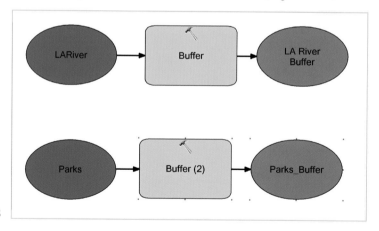

Figure 7-18

4) **Erase park buffers.** Now, as in Lesson 6, we'll use the Erase tool to subtract the park buffers from the LA River buffer.

A Position the ModelBuilder window so you can see the right side of the ArcMap application window.

▷ If necessary, make the Model window smaller or slide it partway off the computer screen.

ⓑ In ArcMap, open the Search window.

 ▷ If the Search tab isn't there, click the Search button 🖥 on the Standard toolbar in ArcMap.

ⓒ Near the top of the Search window, click Tools to limit the search to tools.

ⓓ In the Search box, type **erase** and press Enter.

 ▷ Or pick "erase (analysis)" from the drop-down list.

The Erase (analysis) tool is returned at the top of the list of tools.

ⓔ Hover over the tool name, as in Figure 7-19.

Figure 7-19

ⓕ Drag and drop the tool into the Model window to the right of the output data elements (Figure 7-20).

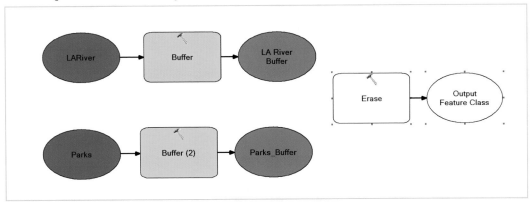

Figure 7-20

ⓖ Double-click the Erase tool to open it.

ⓗ Set the input features to *LA River Buffer* and the erase features to *Parks_Buffer*.

I Compare your Erase dialog box to Figure 7-21 and click OK.

Figure 7-21

In the model, the output datasets of the Buffer tools are now connected as inputs to the Erase tool.

The name of the Erase tool output is long. We don't care about the feature class name, because we don't plan to save it, but we want the name to look better in the model.

J Right-click the *LARiverBuffer_Erase* element and choose Rename.

K Rename the element **Proximity Zone** (this is similar to its name in Lesson 6) and click OK (Figure 7-22).

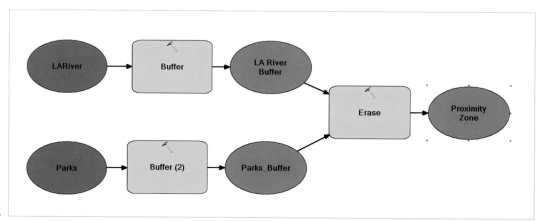

Figure 7-22

ⓛ Right-click the *Proximity Zone* element again to see its context menu.

Notice that the Intermediate setting is unavailable (Figure 7-23).

ⓜ Click in some empty white space to close the context menu and unselect the element.

ⓝ Right-click any blue input data element to see that its Intermediate setting is unavailable, too.

ⓞ Right-click the *Parks_Buffer* element and check its status.

A check mark flags this element as Intermediate.

ⓟ Right-click the *LA River Buffer* element.

It's also flagged as Intermediate. Since this dataset is one we want to keep in the output geodatabase, we'll unflag it.

ⓠ On the context menu, click Intermediate to uncheck it.

▷ Optionally, open the context menu again to confirm that the check mark is gone.

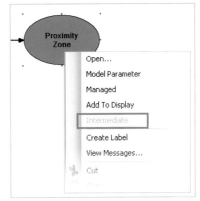

Figure 7-23

Intermediate data

Every data element in ModelBuilder is either intermediate or not. Intermediate data is data that ModelBuilder assumes you don't ultimately want to keep. Input data elements (blue ellipses) can't be intermediate. Output data elements (green ellipses) that come at the end of a chain can't be intermediate either. All other output data elements are flagged as intermediate by default. When you run a model as a tool (Exercise 7D), intermediate data is deleted automatically and never appears in the output geodatabase. When you run a model from the Model window, intermediate data is still written to the output geodatabase, but can subsequently be deleted with the Delete Intermediate Data command on the Model menu. Any data element that is flagged as intermediate can be unflagged if you want to save it.

5) Assign block group demographics to the proximity zone. After erasing the park buffers in Lesson 6, we used the Identity tool to overlay the remaining area of interest with block group attributes (Exercise 6b). We'll do the same thing now in the model.

ⓐ In the ArcMap Search window, delete the existing search term, type **identity**, then press Enter.

ⓑ From the list of results, drag and drop Identity (Analysis) into the Model window near the Proximity Zone output element.

ⓒ Open the Identity tool. Set the Input Features to *Proximity Zone* and the Identity Features to *BlockGroups*.

ⓓ Compare your Identity tool dialog box to Figure 7-24, then click OK.

Figure 7-24

The Identity tool is colored in and connected to its two inputs. At this point, the model layout is getting complicated. We'll adjust it after we rename the output element of the Identity tool.

ⓔ Rename LARiverBuffer_Erase_Identity to **Proximity Zone Identity**. Click OK.

ⓕ Unselect the output data element by clicking in white space.

ⓖ On the ModelBuilder toolbar, click the Auto Layout button ▦, then click the Full Extent button ⬚.

Your layout should look pretty much like Figure 7-25.

ⓗ Save the model.

Figure 7-25

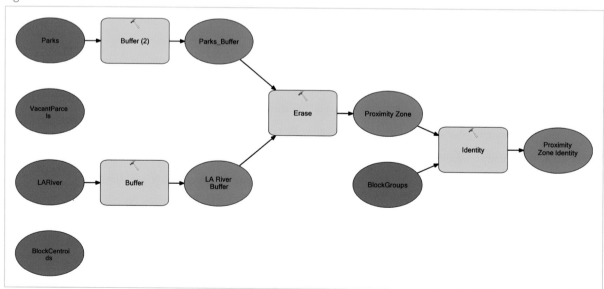

6) Validate and run the model. This is a good point to test-run the model. Before running a model, you should validate it to make sure that all is well with the data variables and tool parameters.

ModelBuilder will write the output data to the ModelBuilderOutputs geodatabase. By default, it doesn't add layers to ArcMap. If we want to see the data, we have to tell ModelBuilder to add it (or add it from the geodatabase later).

❶ In the Model window, right-click the *LA River Buffer* output data element and choose Add To Display (Figure 7-26).

❷ In the same way, add *Proximity Zone Identity* (the last output in the model) to the display.

❸ On the ModelBuilder toolbar, click the Validate Entire Model button ✔.

Figure 7-26

Validating models

Validation ensures that a model has access to its source data. If you rename, move, or delete a feature class referenced by an input data variable, that variable and its dependent processes will turn white (not ready) when you validate the model. Similarly, if you delete a field from a table, and that field is specified by a tool's parameter, the tool and its dependent processes will turn white when you validate.

The next step is to run the model. As it runs, you should get a visual progress report, with each tool turning red as it is processed. (If you don't see the red highlight, it's okay—the job is still being done.)

❹ On the ModelBuilder toolbar, click the Run button ▶.

When the model finishes running, you get a message box (Figure 7-27).

❺ Close the message box.

Scroll through this box to see the order in which processes were run, at what times, and so on.

Figure 7-27

Figure 7-28

Figure 7-29

In the Model window, the tools and output data elements have drop shadows. This tells you that the model has been run.

F Minimize the Model window or move it away from the ArcMap window.

The two outputs marked Add To Display should appear as layers in ArcMap (Figure 7-28).

(If they don't show up, you can add them from the ModelBuilderOutputs geodatabase. The feature class names are *LARiverBuffer* and *LARiverBuffer_Erase_Identity*. You can then rename the layers as shown in Figure 7-28.)

G In the Catalog window, under AnalysisData, expand the ModelBuilderOutputs geodatabase.

It contains four feature classes (Figure 7-29). They have the names assigned by the tool parameters in ModelBuilder. The ArcMap layers have the display names given to the output data elements.

H In the Table of Contents, open the attribute table for the *Proximity Zone Identity* layer. Scroll across to see its attributes.

The table has 285 records and includes the demographic attributes from the *BlockGroups* feature class. It's the same result that we got in Exercise 6b.

I Close the attribute table.

J In the Table of Contents, remove the *Proximity Zone Identity* and *LA River Buffer* layers from the map.

K Restore the Model window, or move it back into the center of your screen.

L Save the model.

We'll finish the model in the next exercise. If you want, you can take a break here.

M Optionally, from the ModelBuilder main menu, choose Model → Close.

N Optionally, exit ArcMap. If you exit, don't save your changes.

Adding layers to the display

After the model has been run, you can right-click any output data element in the Model window and choose Add To Display to add it as a layer to ArcMap.

Exercise 7c: Build the model (Part 2)

In this exercise, we'll keep following the Lesson 6 workflow and finish the analysis model. As mentioned, we're not going to aggregate the three candidate parcels that are close to each other, so our results won't be identical to Lesson 6.

In this exercise, we'll see how ModelBuilder handles operations on layers. In Lesson 6, we used attribute and spatial queries to narrow down the number of potential park sites. Queries (and, for that matter, any type of feature or record selections) are not made directly on feature classes, but rather on layers. To make feature selections in ModelBuilder, we therefore need a special tool that turns an input feature class into a layer. This tool is called Make Feature Layer. When we make layers in ModelBuilder, it's usually for the purpose of making a selection that can be used as an input to further processes. For a specialized use of the Make Feature Layer tool in ArcMap, see the topic *Apportioning attribute values*.

1) **Select features by demographic attributes.** In Lesson 6, we used an attribute query to select features in the proximity zone according to demographic values. We'll re-create that query now.

Ⓐ If necessary, start ArcMap. In the Getting Started dialog box, open Lesson7 from the list of recent maps.

▷ If ArcMap and ModelBuilder are still open from the previous exercise, skip to Step 1C.

Ⓑ In the Catalog window, in the Home folder, expand the ParkPlanningModels toolbox. Right-click *Park Suitability Analysis* and choose Edit.

Ⓒ Move the Model window far enough out of the way that you can use the ArcMap Search window.

Ⓓ In the ArcMap Search window, click Tools, if necessary, and search for **select by attribute**.

In the list of returned tools, Select Layer By Attribute (Data Management) is the one we want (Figure 7-30). The tool's name tells us that it needs to work on a layer.

Ⓔ Drag and drop Select Layer By Attribute into the Model window. Don't worry about where it goes exactly.

Ⓕ Double-click the tool to open it.

Figure 7-30

Figure 7-31

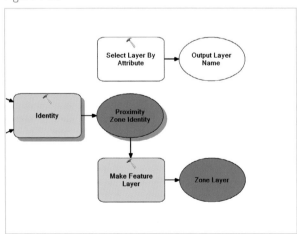

Figure 7-33

In the tool dialog box, there's no drop-down list for the required input (Layer Name or Table View) because there aren't any layers or tables in the ArcMap document.

G Close the Select Layer By Attribute dialog box.

H In the ArcMap Search window, search for `make feature layer`.

In the list of returned tools, we want Make Feature Layer (Data Management) (Figure 7-31).

I Drag and drop Make Feature Layer into the Model window and open the tool.

J Set the Input Features to *Proximity Zone Identity*.

K Replace the Output Layer path with `Zone Layer` (Figure 7-32) and click OK.

Figure 7-32

The tool is connected to the *Proximity Zone Identity* element and filled in.

The layout may be different, but the elements in your Model window should be connected as in Figure 7-33.

L Right-click the *Proximity Zone Identity* element.

We're not going to look at it again on the map, so we'll unflag its display property.

M Uncheck Add To Display.

N Open the Select Layer By Attribute tool.

O Set the Layer Name or Table View to *Zone Layer*.

This choice is available because you defined *Zone Layer* with the Make Feature Layer tool. *Zone Layer* is not a dataset: it's a layer made from the *Proximity Zone Identity* output—which actually means a layer that references the *LARiverBuffer_Erase_Identity* feature class.

P Next to the Expression box, click the SQL button to open the Query Builder.

ⓠ Build the expression shown in Figure 7-34:

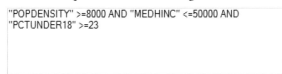

```
"POPDENSITY" >=8000 AND "MEDHINC" <=50000 AND
"PCTUNDER18" >=23
```

Figure 7-34

ⓡ Click OK on the Query Builder dialog box. Confirm that the Select Layer By Attribute dialog box looks like Figure 7-35 and click OK.

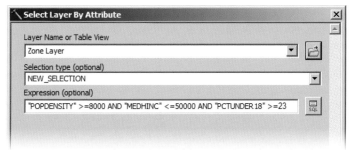

Figure 7-35

ⓢ Rename the output data element from Zone Layer (2) to **Good Zones Layer**.

ⓣ On the ModelBuilder toolbar, click the Auto Layout button ▮▮, and then the Full Extent button ⊡.

The model may be too big now for you to make out the names of individual elements when you're zoomed to the full extent.

ⓤ On the ModelBuilder toolbar, click the Zoom In tool ⊕ and zoom in on the right side of the model.

ⓥ From the ModelBuilder main menu, choose View ➝ 75% (Figure 7-36).

Figure 7-36

ⓦ Save the model.

Zooming and panning your model

The ModelBuilder toolbar has several navigation tools for zooming in and out on your model: ⊡ ▨ ▨ ⊕ and ✋. You can also use the horizontal and vertical scroll bars. The View menu includes some of the same tools as well as predefined and custom zoom level choices.

Figure 7-37

2) Select parcels within good zones. The attribute query for desirable demographics significantly narrows down the "good zones" where we can look for vacant parcels. We now have the ModelBuilder equivalent of an attribute selection on a layer, bringing us to the point in the analysis where we were at the end of Lesson 6b. The next thing to do is to make a spatial query for vacant parcels that lie within the good zones.

Ⓐ In the ArcMap Search window, search for `select by location`.

In the list of returned tools, the one we want is Select Layer By Location (Data Management) (Figure 7-37).

Ⓑ Drag and drop Select Layer By Location into the Model window.

Since this tool also requires a layer as input, we need to make a feature layer from the *VacantParcels* input element.

Ⓒ In the ArcMap Search window, search for `make feature layer`. Drag and drop the Make Feature Layer tool into the model.

Ⓓ Open the Make Feature Layer (2) tool and set the Input Features to *VacantParcels*.

In Lesson 6 we did a spatial query for vacant parcels in good zones, followed by an attribute query for parcels one acre or larger. We can shortcut the attribute query on parcel size by incorporating the query in the Expression parameter of the Make Feature Layer tool. (We actually could have done the same thing with the attribute query in the previous step, too.)

Building the query in the Expression parameter of the Make Feature Layer tool switches the order of the two query operations from Lesson 6, but doesn't affect the logic. It doesn't matter whether we first select the vacant parcels in good zones and then subselect the one-acre ones (Lesson 6) or first select all the one-acre vacant parcels and then subselect the ones in good zones (as we'll do here).

Ⓔ Next to the Expression box, click the SQL button.

Ⓕ In the Query Builder, create the expression shown in Figure 7-38 and click OK.

Figure 7-38

G Replace the default Output Layer name with **Vacant Parcels 1 Acre Layer**.

Figure 7-39

H Confirm that the dialog box looks like Figure 7-39 and click OK.

I Click Auto Layout and zoom to the full extent to see how the model has been rearranged.

J Zoom in to a more legible level, such as 75% or 100%.

K Optionally, rename any elements with bad line breaks by inserting spaces in their names.

▷ For example, you might rename the *VacantParcels* input data element to **Vacant Parcels**.

The part of the model that you're working with should look like Figure 7-40.

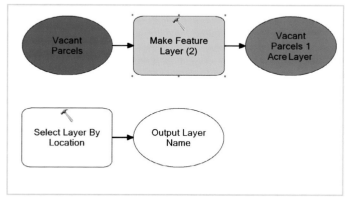

Figure 7-40

L Open the Select Layer By Location tool and set the Input Feature Layer to *Vacant Parcels 1 Acre Layer.*

M If necessary, set the Relationship to INTERSECT.

N Set the Selecting Features to *Good Zones Layer.*

Vacant one-acre parcels will be selected if they intersect good zones.

O Make sure your Select Layer By Location dialog box looks like Figure 7-41 and click OK.

Figure 7-41

P Click the Auto Layout button and scroll to the right side of the Model window.

Q Rename the last output data element to **Good Zone Parcels Layer** (Figure 7-42).

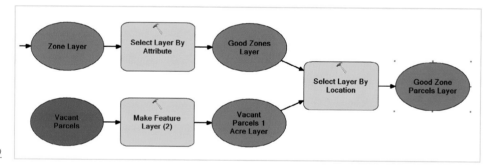

Figure 7-42

Note that the last output that actually creates a dataset is *Proximity Zone Identity.* The subsequent outputs consist of two layers created by the Make Feature Layer tool and selections on those layers.

R Hover over the Proximity Zone Identity element.

The ToolTip shows the path to the dataset on disk (Figure 7-43).

Figure 7-43

S Hover over each of the downstream outputs in turn (Figure 7-44).

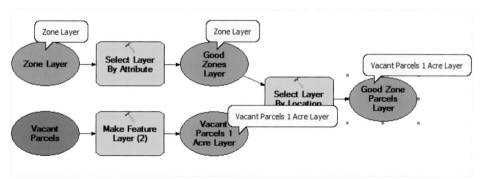

Figure 7-44

These ToolTips show only the name of the layer, because that's all there is to see—no datasets have been created on disk. There are four output elements, but only two layers: *Zone Layer*, created by Make Feature Layer, and *Vacant Parcels 1 Acre Layer*, created by Make Feature Layer (2). *Good Zones Layer* is an attribute selection on *Zone Layer*, which is why its ToolTip shows the same layer name. The same is true for *Good Zone Parcels Layer*: it's a spatial selection on *Vacant Parcels 1 Acre Layer*. We renamed the output elements of the selection tools for clarity (hopefully), but we didn't have to.

3) Copy features. In Lesson 6, we used the Export Data command to write our selection to a feature class, but in ModelBuilder we have to rely on geoprocessing tools.

A In the ArcMap Search window, locate the Copy Features (Data Management) tool. Drag and drop it into the Model window.

B Open the tool and set the Input Features to *Good Zone Parcels Layer*.

C Replace the Output Feature Class path with `RecommendedSites` and press Tab.

D Make sure the Copy Features dialog box looks like Figure 7-45 and click OK.

E Click the Auto Layout button and the Full Extent button.

F Save the model.

Figure 7-45

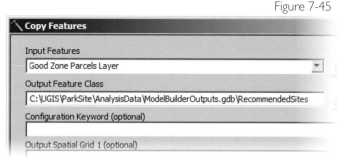

At this point in Lesson 6, we aggregated the three sites that were right by each other. Since we're not doing that here, we'll move on to create access zones around the candidates and count the total population inside them.

4) Create park access zones. The access zones are quarter-mile buffers. Rather than add the Buffer tool again from the Search window, we'll copy and paste it from inside the model.

Ⓐ Zoom in on the left side of the model.

Ⓑ Right-click the Buffer (2) tool and choose Copy.

Ⓒ Right-click in some empty white space in the Model window and choose Paste.

Ⓓ Click Auto Layout (Figure 7-46).

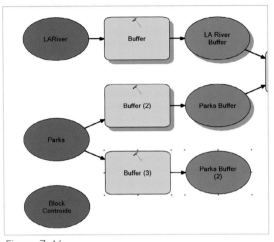

Figure 7-46

A Buffer (3) tool is created. Like Buffer (2), it's connected to the *Parks* input data. All the parameters of the Buffer (2) tool have been copied as well.

Ⓔ Open the Buffer (3) tool and set the Input Features to *RecommendedSites*.

The yellow icon ⚠ by the Output Feature Class parameter warns you that another feature class already has the name *Parks_Buffer*.

Ⓕ Replace the Output Feature Class path with **ParkAccessZones** and press Tab.

Ⓖ Keep the same distance value (0.25 miles) but change the Dissolve Type to NONE.

Ⓗ Make sure the Buffer (3) dialog box looks like Figure 7-47 and click OK.

Ⓘ Click Auto Layout and scroll to the right side of the Model window.

Ⓙ Rename the output element of the Buffer (3) tool to **Park Access Zones**.

5) Count total population within park access zones. To get a population count, we'll spatially join *BlockCentroids* to *ParkAccessZones*. The output will have the same features as *ParkAccessZones* and the attributes of both inputs. A merge rule will sum the populations of the block centroids within each access zone.

Ⓐ In the ArcMap Search window, search for **spatial join**.

Ⓑ Drag and drop the Spatial Join (Analysis) tool into the Model window.

Ⓒ Open the tool and set the Target Features to *Park Access Zones*.

Ⓓ Set the Join Features to *Block Centroids*.

Figure 7-47

ⓔ Accept the default Output Feature Class path.

We'll delete unnecessary fields from the output attribute table. Delete carefully so that your mouse clicks don't get ahead of you.

ⓕ In the Field Map of Join Features area, click the first field in the list (it should be SHAPE_LENGTH) to select it.

ⓖ Click the Delete button ✖ to delete the field.

ⓗ Go on to delete all the fields except ACRES (now second from the top) and POP2010 (third from the bottom).

  ▷ If you accidentally delete either of these two fields, close the tool and start again.

ⓘ Right-click the POP2010 field and choose Merge Rule → Sum.

Underneath the Field Map of Join Features box is the Match Option parameter, which defines the spatial relationship to be evaluated. Either INTERSECT (which includes containment) or CONTAINS will work.

ⓙ Accept the default Match Option setting of INTERSECT.

ⓚ Compare your dialog box to Figure 7-48 and click OK.

ⓛ Rename the output element of the Spatial Join tool to **Park Access Zones Pop**.

ⓜ Do an auto layout and zoom to the full extent.

The model is essentially done. We'll skip the distance calculation to the river, as well as the overlays that brought the demographic attributes into *RecommendedSites* in Lesson 6. The last thing we'll do before running the model is set the status of our output data elements correctly. The three outputs that we want to save to the geodatabase and add as layers to ArcMap are:

- *LA River Buffer*
- *Recommended Sites*
- *Park Access Zones Pop*

All other outputs are intermediate data.

ⓝ Right-click each green output data element and confirm or set its Add To Display and Intermediate status as shown in Figure 7-49.

Figure 7-48

Figure 7-49

Output Element	Add To Display	Intermediate
Parks Buffer		√
LA River Buffer	√	
Proximity Zone		√
Proximity Zone Identity		√
Zone Layer		√
Vacant Parcels 1 Acre Layer		√
Good Zones Layer		grayed-out ←
Good Zone Parcels Layer		grayed-out ←
Recommended Sites	√	
Park Access Zones		√
Park Access Zones Pop	√	grayed-out

Grayed-out because they are selections on a layer

⓪ Save the model and zoom to the full extent.

6) **Validate and run the model.** We're ready to run the model.

Ⓐ On the ModelBuilder toolbar, click the Validate Entire Model button ✔.

This removes drop shadows from elements that have already been run and returns them to their ready-to-run state.

Ⓑ Click the Run button ▶.

Ⓒ When the model finishes, close the dialog box. Minimize the Model window or move it out of the way.

7) **View results.** If the model ran successfully, you should see the *Park Access Zones Pop, Recommended Sites,* and *LA River Buffer* layers in ArcMap.

　▷ If you don't see these layers, restore the Model window. Check or uncheck Add To Display for output data elements as needed.

Ⓐ Drag *Recommended Sites* to the top of the Table of Contents.

　▷ If necessary, click the List By Drawing Order button at the top of the Table of Contents.

Ⓑ Optionally, resymbolize the layers.

Ⓒ Open the *Park Access Zones Pop* attribute table (Figure 7-50).

OBJECTID *	Shape *	Join_Count	TARGET_FID	ACRES	POP2010	Shape_Len
1	Polygon	26	1	1.494605	3113	9560.1
2	Polygon	26	2	2.362151	3070	9872.47
3	Polygon	22	3	1.243734	2251	9302.45
4	Polygon	12	4	1.658465	1475	9447.83
5	Polygon	13	5	3.433686	1338	9963
6	Polygon	9	6	1.950284	3107	9599.38
7	Polygon	26	7	1.711671	3820	9720.52

Figure 7-50

The table has seven records instead of five because of the three sites that weren't aggregated. The values in the ACRES and POP2010 fields should match your Lesson 6 results for four of the records.

Ⓓ Close the table.

Ⓔ Remove all layers from the Table of Contents.

Ⓕ In the Catalog window, expand the ModelBuilderOutputs geodatabase, if necessary.

It should contain seven feature classes.

Figure 7-51

G Restore the Model window. From the main menu, choose Model → Delete Intermediate Data.

Data marked Intermediate in the model is deleted from the geodatabase. You should be left with the three feature classes shown in Figure 7-51.

H Save and close the model.

I Save the map document. Optionally, exit ArcMap.

Troubleshooting a model

If a model process fails or a warning is issued, you'll see it in the message box. Warnings appear with a code that links to a description in the ArcGIS Desktop Help. One way to check whether individual outputs are okay is to add them to the display so you can see the layer in ArcMap. If you combine this technique with incremental building and testing of the model, you can save yourself a lot of backtracking when things go wrong. If you add processes to a model and rerun it without validating the entire model, only the downstream part (the part without drop shadows) will be executed.

Exercise 7d: Run the model as a tool

In this exercise, we'll set up the model as a geoprocessing tool. This is an efficient way to run a model, and it's a good way to share the model with people who may want to use it but are not so interested in the details of its construction. When you run a model as a tool, it has a dialog box interface like any other geoprocessing tool: the user sets a number of parameters and clicks OK. Behind the scenes, of course, the model is a long chain of tools, each of which has its own required and optional parameters: if all these parameters had to be set by the user in the model's dialog box, it would be a long process.

Fortunately, that's not how it works. When you set up the model to run as a tool, you specify which parameters from which tools inside the model will be exposed in the model dialog box. Only these parameters are filled out by the user (you can supply default values): all others are preset by you, and the person running the tool never sees them. In other words, you decide how flexible to make the tool.

The ability to test sensitivity to changes in variables and to explore alternative outcomes are key reasons for building models in the first place. Accordingly, when we set up the model to run as a tool, we should think about which factors the user is most likely to want to experiment with.

1) **Open the model.** In previous exercises, you've *edited* the model, which is how you start work in the Model window. To run the model as a geoprocessing tool, you *open* it.

Ⓐ If necessary, start ArcMap and open Lesson7 from the list of recent maps.

Ⓑ In the Catalog window, in the Home folder, expand the ParkPlanningModels toolbox. Right-click *Park Suitability Analysis* and choose Open.

The model opens as a geoprocessing tool (Figure 7-52). It's just an empty gray dialog box with the message, "This tool has no parameters." That's because we haven't yet gone into the Model window and chosen the parameters we want to expose.

Figure 7-52

If the tool's help panel is showing, you'll see the text you entered in the Model Properties in Exercise 7a.

Ⓒ Close the Park Suitability Analysis dialog box.

Ⓓ In the Catalog window, right-click the model and choose Edit.

Ⓔ In the Model window, zoom in on the left side of the model and set your view scale to a comfortable level.

▹ How? Click the View menu and choose a value.

2) **Make tool parameters for buffer size.** Users of the model tool will probably want to experiment with the buffer sizes for maximum distance from the river (0.75 miles) and minimum distance from existing parks (0.25 miles). To expose these parameters to the tool's dialog box, we first have to define them as variables. This allows their values to be changed any time the model is run. Then we flag the variables as "model parameters" (parameters we want to include in the model's tool dialog box).

Ⓐ In the Model window, right-click the Buffer tool connected to the *LARiver* input, and choose Make Variable → From Parameter → Distance [value or field].

This says that we want the distance parameter in the Buffer tool to be a variable. A small light blue ellipse, called a value variable, is added to the Model window.

❸ On the ModelBuilder toolbar, click the Select tool ↖, if necessary. Click once on the value variable so that it remains selected while the Buffer tool is unselected.

❹ Drag the value variable slightly away from the *LARiver* element.

❺ Right-click the variable and choose Open (Figure 7-53).

Figure 7-53

Within the Model window, the Distance parameter is now detached from the Buffer tool and can only be set here. We'll accept the current value of 0.75 miles as the default for the model tool.

❻ Close the Distance dialog box.

❼ Right-click the value variable and choose Model Parameter.

The letter "P" (for "parameter") appears above the variable (Figure 7-54).

In the model tool dialog box, the title of the parameter will match the variable name. We don't really want the parameter to be called "Distance [value or field]."

❽ Right-click the value variable and choose Rename. For the new name, type **Distance from LA River** (Figure 7-55) and click OK.

Figure 7-55

Figure 7-54

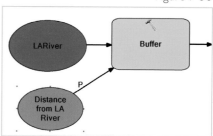

The variable is flagged as a model parameter.

We'll resize the element so we can read the text.

❾ With the value variable selected, click and drag one of its corner selection handles until you can see the whole name (Figure 7-56).

Figure 7-56

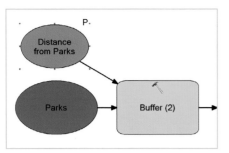

Figure 7-57

We'll make a parameter for the *Parks* buffer distance in the same way.

❶ Right-click the Buffer (2) tool connected to the *Parks* input, and choose Make Variable ➞ From Parameter ➞ Distance [value or field].

❿ Click on the new variable so that it's the only selected element and move it off the *Parks* input element.

Ⓚ Right-click the variable and choose Model Parameter.

Ⓛ Rename the variable **Distance from Parks** and resize it as needed (Figure 7-57).

3)   Make parameters for demographics and park size. We'll make tool parameters from the query expressions for demographics and park size so we can experiment with these, too.

Ⓐ Scroll to the middle of the model.

Ⓑ Right-click the Select Layer By Attribute tool and choose Make Variable ➞ From Parameter ➞ Expression.

A value variable with the name Expression is added to the Model window.

Ⓒ Make the new variable the only selected element and move it to a good place.

Ⓓ Hover over the variable element to see the contents of the expression as a tip (Figure 7-58).

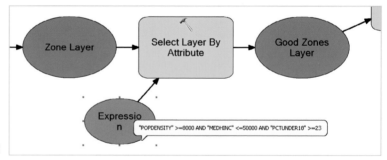

Figure 7-58

Ⓔ Flag the variable as a model parameter.

Ⓕ Rename the variable **Demographic Criteria** and resize it as needed (Figure 7-59).

▷ You can drag a side selection handle to widen the ellipse without changing its height.

Ⓖ Right-click the Make Feature Layer (2) tool connected to the *Vacant Parcels* input, and choose Make Variable ➞ From Parameter ➞ Expression.

Ⓗ Make the new variable the only selected element and move it to a good place.

Figure 7-59

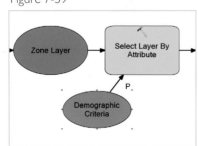

- **ⓘ** Hover over the element to see its expression ("ACRES" >= 1).
- **ⓙ** Flag the variable as a model parameter.
- **ⓚ** Rename the variable **Park Size in Acres** and resize it as needed (Figure 7-60).
- **ⓛ** Click the Validate Entire Model button ✔.
- **ⓜ** Save and close the model.

4) **Run the tool.** Now we'll reopen the model as a tool and run it.

- **ⓐ** In the Catalog window, right-click *Park Suitability Analysis* and choose Open.

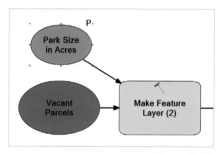

Figure 7-60

The tool displays the parameters you've specified (Figure 7-61). We'll experiment with new values on the run after this one. This time, we'll run the model with default values to make sure it works.

- **ⓑ** Click OK on the Park Suitability Analysis dialog box.
- **ⓒ** When the model is finished running, close the message box.

The model gives the same results whether you run it from ModelBuilder or as a tool, but its execution is a little different. For one thing, when the model is run as a tool, intermediate data is deleted automatically.

- **ⓓ** In the Catalog window, expand, if necessary, the ModelBuilderOutputs geodatabase.

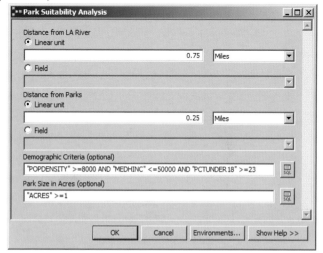

Figure 7-61

There should be three feature classes:

- *LARiverBuffer*
- *ParkAccessZones_SpatialJoin*
- *RecommendedSites*

By default, ArcMap doesn't allow geoprocessing outputs to overwrite each other, yet these three feature classes replaced three of the same name that were there before. Also, by default, ArcMap *does* add geoprocessing results as layers to the map, but no layers were added here. As we'll see next, the model's output datasets also have to be exposed as tool parameters for these geoprocessing defaults to take effect.

Reordering tool parameters

To change the order in which parameters are displayed in the dialog box, open the Model Properties and click the Parameters tab. The parameters appear in a list where they can be selected and moved up or down with arrow keys. The Model Properties can be opened from the Model menu in ModelBuilder, or from the model's context menu in the Catalog window.

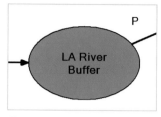

Figure 7-62

5) Make tool parameters for the outputs. Flagging the outputs of the model as tool parameters will let us add the results as layers to ArcMap. More importantly, it will give us control over whether output datasets overwrite each other. As we experiment with different values in successive runs of the model, we may want to save our results with different names so we can compare them.

A In the Catalog window, right-click *Park Suitability Analysis* and choose Edit.

B In the Model window, right-click the *LA River Buffer* output data element and choose Model Parameter (Figure 7-62).

We don't have to make a variable first, because data elements are variables by definition.

C Right-click the element again to open its context menu.

The Intermediate status, previously unchecked, is now unavailable. Once output data is made a model parameter, it can't be intermediate.

D Click in some empty white space to close the context menu.

E Set *Recommended Sites* and *Park Access Zones Pop* as model parameters.

F Validate the model, save it, and close it.

G In ArcMap, from the main menu, choose Geoprocessing → Geoprocessing Options.

Figure 7-63

H In the Geoprocessing Options dialog box, make sure that "Overwrite the outputs of geoprocessing operations" is unchecked.

I Make sure that "Add results of geoprocessing operations to the display" is checked (Figure 7-63) and close the dialog box.

Now that the outputs are tool parameters, they will not be allowed to overwrite feature classes of the same name in the geodatabase. (They won't be allowed to do it when the model is run as a tool. When the model is run from the Model window, it isn't governed by these geoprocessing settings.)

6) Run the tool with different values. Now we'll run the model with a new river buffer value.

A In the Catalog window, open the model to run as a tool.

▷ You can also do this by double-clicking its name.

The three new parameters you added show up in the tool dialog box (Figure 7-64). They are marked with error icons ⊗ because feature classes with these names already exist in the geodatabase. You can't run the tool until you fix the errors.

Figure 7-64

If ArcMap were set to overwrite inputs in the Geoprocessing Options, you would see a warning icon ⚠ next to these parameters instead. The warning would alert you to change the output names, but you could still run the tool without doing so.

Ⓑ Change the *LARiverBuffer* output feature class name to **LARiverBuffer2**.

Ⓒ Change the *RecommendedSites* output feature class name to **RecommendedSites2**.

Ⓓ Change the *ParkAccessZones_SpatialJoin* output feature class name to **ParkAccessZones2**.

As you click out of each parameter, its warning icon goes away.

Ⓔ Finally, change the Distance from LA River value to 0.25 miles.

Ⓕ Compare your dialog box to Figure 7-65 and click OK.

Ⓖ When the model finishes, close the message box.

You should see three layers in ArcMap. The LA River buffer is skinnier than usual: just a half-mile wide now, instead of a mile and a half.

Ⓗ In ArcMap, drag *RecommendedSites2* to the top of the Table of Contents. If necessary, drag *LARiverBuffer2* to the bottom.

Ⓘ Optionally, resymbolize the layers and add *LARiver* from the ReadyData geodatabase.

Ⓙ Zoom in on the suitable site (Figure 7-66).

Figure 7-65

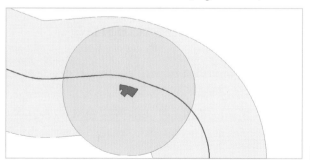

Figure 7-66

Using 0.25 miles as the value for distance to the river, we only find one candidate.

K In the Catalog window, expand ModelBuilderOutputs, if necessary.

The geodatabase should contain six feature classes: three from this run of the model and three from the previous run.

L Save the map document. Optionally, exit ArcMap.

Exercise 7e: Document and share the model

If you want people to understand how your model works, you should invest some time in documenting it. The item description is a good place to explain how the model works as a geoprocessing tool. Within the Model window, we can add notes, named labels, to explain what the model is doing at any given point. We can also change the layout to make the model easier to read.

Eventually, you may want to share your model with colleagues so they can review it, improve it, or adapt it for their own purposes. You share a model by exporting it as a package from the geoprocessing results, which we looked at previously in Lesson 4.

1) **Edit the item description.** Like datasets, tools have item descriptions. A tool's item description can summarize its purpose and provide usage notes. It can also include keyword tags for finding the tool in searches.

A If necessary, start ArcMap and open Lesson7.

B In the Catalog window, in the Home folder, expand the ParkPlanningModels toolbox. Right-click the model and choose Item Description.

C At the top of the Item Description window, click Edit (Figure 7-67).

The Title has already been filled in. We'll skip the thumbnail image. Scroll down, or resize the Item Description, as needed, to complete the next steps.

D In the Tags box, enter some useful words and phrases, such as `parks, rivers, site selection, suitability analysis, Los Angeles`.

The Summary has already been filled in with the model description that you provided in Exercise 7a.

The Usage box lets you document things about the tool's workings that might not be self-evident. In this case, there isn't anything especially urgent to say.

Figure 7-67

ⓔ Optionally, in the Usage box, describe how to use the tool: **Choose a maximum distance from the river, a minimum distance from existing parks, desired neighborhood demographics, and parcel size in acres to find suitable parcels.**

ⓕ Under Syntax, expand the Distance_from_LA_River parameter.

ⓖ In the Dialog Explanation box, enter **The maximum distance a suitable site can be from the river.** (Figure 7-68)

ⓗ Expand the Distance_from_Parks parameter.

ⓘ In the Dialog Explanation box, enter **The minimum distance a suitable site can be from an existing park.**

ⓙ Add your own dialog explanations for the other parameters, including the output dataset parameters.

It's important that each parameter have an explanation, as you'll see later in the exercise.

ⓚ Near the bottom of the Item Description, in the Credits box, enter **City of Los Angeles, Esri.**

ⓛ Scroll to the top of the Item Description and click the Save button (Figure 7-69).

ⓜ Close the Item Description.

Distance_from_LA_River
Dialog Explanation

B *I* U A⁺ A⁻ ☰ ☱ ✎ ☰ ☰ ☰

The maximum distance a suitable site can be from the river.

Figure 7-68

Figure 7-69

Item Description - Park Suitability Analysis

Description

🖶 Print   📝 Edit

Park Suitability Analysis

Title   Park Suitability Analysis

Summary
This model identifies vacant parcels near the Los Angeles River that may be suitable for a new park.

Usage
Choose a maximum distance from the river, a minimum distance from existing parks, desired neighborhood demographics, and parcel size in acres to find suitable parcels.

Syntax
ParkSuitabilityAnalysis (Distance_from_LA_River, Distance_from_Parks, {Demographic_Criteria}, {Park_Size_in_Acres}, LA_River_Buffer, RecommendedSites, Park_Access_Zones_Pop)

2) Change the model layout. The orientation of the model layout can be changed in various ways.

ⓐ In the Catalog window, right-click the model and choose Edit.

ⓑ Click the Auto Layout and Full Extent buttons.

ⓒ From the ModelBuilder main menu, choose Model → Diagram Properties.

Figure 7-70

Figure 7-71

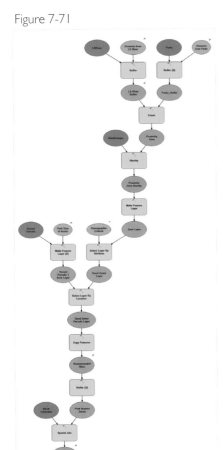

Ⓓ On the Diagram Properties dialog box, click the Layout tab.

Ⓔ In the Orientation area, click Top to Bottom (Figure 7-70) and click OK.

Nothing happens until you click Auto Layout.

Ⓕ Click Auto Layout and then Full Extent (Figure 7-71).

This orientation is better for some purposes, such as printing on a vertically oriented page.

Ⓖ On the ModelBuilder toolbar, click the Undo button ↶.

The layout reverts to a horizontal orientation. (The settings don't revert, though: if you click Auto Layout, the model will flow top to bottom again.)

Ⓗ Open the Diagram Properties again and click the General tab.

Ⓘ In the Layout Mode area, click Manual. Click OK.

The best-looking layout may be one that you design yourself. We'll keep the orientation horizontal, but split it into three tiers. (If you see a way you like better, go ahead and try it.)

Ⓙ On the ModelBuilder toolbar, click the Select tool ▶.

Ⓚ Drag a selection box as shown in Figure 7-72.

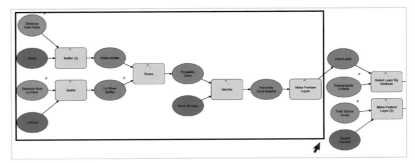

Figure 7-72

When you release the mouse button, the elements covered by the box are selected. You can modify the selection by pressing the Control key and clicking an element to select or unselect it.

Ⓛ Place the pointer over any selected element; then click and drag the entire selection toward the upper left corner of the Model window.

The model remains connected, although a connector arrow may get pretty long.

M Drag another selection box as shown in Figure 7-73.

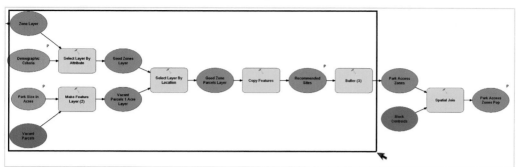

Figure 7-73

N Drag the selected elements underneath the first tier of elements and give them a comfortable indent.

O Drag a box around the last set of elements and move them underneath the second tier, again with an indent.

P Optionally, reshape connector arrows by clicking anywhere on them and dragging from the click point to a new location.

Q Zoom in on the model.

In the best case, the model fits legibly in your Model window. (You might need to enlarge your window.) The result should look something like Figure 7-74.

Figure 7-74

Figure 7-75

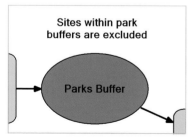

Figure 7-76

R Click Auto Layout.

You get a message that automatic layout is disabled.

> ▷ If you didn't get the message, and the layout changed, click the Undo button on the toolbar. Then open the Diagram Properties, click the General tab, click Manual, and click OK.

S Click OK on the error message and save the model.

3) Label the model. Labels in a model are helpful in the same way as comment lines in a computer program: they explain what's happening. Labels can be attached to specific elements in the model or they can be free-floating.

A From the ModelBuilder main menu, choose View ➔ 100%. Scroll or pan to the upper left corner of the model.

B On the Model toolbar, click the Select tool ▶. Hover over various elements, such as the *Distance from LA River* variable (Figure 7-75).

A lot of documentation is already built into the tool tips. You see the paths of input and output elements, the values of expressions, and the parameters of tools.

C Right-click the *Parks Buffer* output element and choose Create Label.

A label with the default text "Label" is added to the Model window. It may appear on top of a connector arrow.

D Click in empty white space to unselect the label and the output data element; then click the label to select it.

E Drag the label to a position above the output data element.

As you drag the label, it stays connected to its element by a dotted line. The line disappears when you release the mouse button.

F Double-click the label to make it editable.

G Type **Sites within park buffers are excluded.**

H Place the cursor between the words "park" and "buffers" and press Shift-Enter to break the line. Click in any white space to finish the label (Figure 7-76).

I Create labels as suggested below, modifying them as you like:

- For *LA River Buffer:* **Sites outside this buffer are excluded.**
- For *Proximity Zone:* **Only sites in this area are considered.**
- For the Identity tool: **Adds demographic attributes to proximity zone.**

The labels will look better if we change their font properties.

- ⓙ Hold down the Shift key and click all four labels to select them.
- ⓚ Right-click any selected label and choose Display Properties.

The Display Properties window has two columns: on the left are properties that can be set; on the right are the current values of those properties.

- ⓛ In the Display Properties window, click on the value MS Shell Dlg next to the Font property, then click the button that appears next to the font name (Figure 7-77).

Figure 7-77

Click here

- ⓜ In the Font dialog box, change the Font Style from Bold to Regular.
- ⓝ Click the Color drop-down arrow and change the color from Black to Gray.
  - ▷ Or set the font properties to suit yourself.
- ⓞ Compare your dialog box to Figure 7-78 and click OK.
- ⓟ Close the Display Properties window and click in some white space to unselect the labels.

The labels you've created so far are connected to specific model elements. You can also create free-floating labels.

- ⓠ From the ModelBuilder main menu, choose Insert → Create Label.
- ⓡ Drag the label to the top center of the Model window. Change its text to something like **This model finds vacant parcels suitable for park development**.
- ⓢ Open the label's display properties. Make the font style Regular, the color Gray, and the size 18. Click OK.
- ⓣ Close the Display Properties window and unselect the label (Figure 7-79).

Figure 7-78

Figure 7-79

Ⓤ Optionally, add more labels to the model to explain other processes.

Ⓥ Zoom to the full extent.

Setting display properties for model elements

To change the appearance of individual model elements (including connector arrows) or groups of selected elements, right-click an element and choose Display Properties. Among the properties you can set are color, shape, size, and text font.

4) View a model report. It's important to associate the results of your model with the input conditions that led to them. ModelBuilder can formalize this information in a report.

Ⓐ From the ModelBuilder main menu, choose Model → Report.

Ⓑ Click OK to view the report in a window.

The report sorts the model elements into variables and processes (tools).

Ⓒ In the upper right corner of the Model Report window, click the Expand/Collapse All link (Figure 7-80).

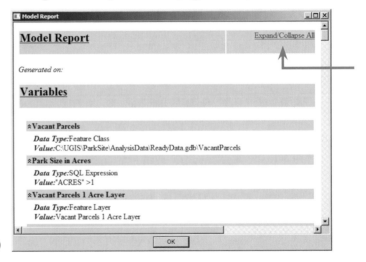

Figure 7-80

Ⓓ Scroll quickly through the report; then click OK on the Model Report window.

By choosing "Save report to a file," instead of "View report in a window," you can save the report to disk as an .xml file.

Ⓔ Save and close the model.

5) **Share the model.** You can share the model by making a geoprocessing package. This is a compressed file containing your model and all its associated input and output data. Once you've made the package, you can send it to a colleague by email or any other file transfer method you want.

Ⓐ From the ArcMap main menu, choose Geoprocessing → Results.

Ⓑ In the Results window, expand Current Session and expand the upper (most recent) of the two Park Suitability Analysis entries.

▷ If you started ArcMap at the beginning of this exercise, you'll need to expand Previous Session instead of Current Session.

Ⓒ Expand the Inputs heading (Figure 7-81).

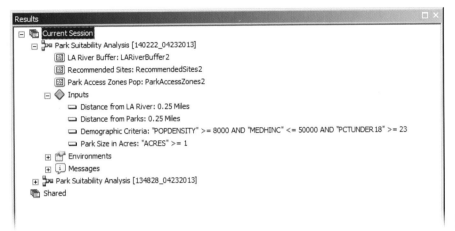

Figure 7-81

The Results window shows the output feature classes and parameter settings from the last time you ran the model as a geoprocessing tool.

Ⓓ In the Results window, right-click the Park Suitability Analysis heading and choose Share As → Geoprocessing Package (Figure 7-82).

Figure 7-82

E In the Geoprocessing Package dialog box, click the Save package to file option.

F Confirm that the output path goes to your MapsAndMore folder and that the file name is ParkSuitabilityAnalysis.gpk

G At the top of the dialog box, click Analyze to check for any problems (Figure 7-83).

Figure 7-83

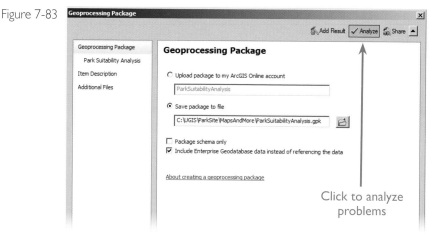

Click to analyze problems

A Prepare window opens in ArcMap to show any problems. Everything should be fine, with no errors, warnings, or even messages, as in Figure 7-84.

Figure 7-84

The reason you needed to provide all those parameter explanations in the Item Description earlier in the exercise was to avoid having them show up as errors here. If you do have any errors, you have to fix them before you can create the geoprocessing package. Warnings or messages may be important but will not prevent packaging.

H At the top of the Geoprocessing Package dialog box, click Share.

I When the package has been successfully created, click OK on the message.

❿ In the Catalog window, expand the ParkSite\MapsAndMore folder, if necessary (Figure 7-85).

Figure 7-85

The package, consisting of the model and its data can now be shared with colleagues who have ArcGIS for Desktop 10.1. (It can't be used with earlier versions.)

Ⓚ Close the Results window.

Using the geoprocessing package

You can open the geoprocessing package by double-clicking it in the Catalog window or Windows Explorer, or by right-clicking it in the Catalog window and choosing Unpack. You can also drag and drop the package onto the ArcMap window.

When the package opens, the effect is as though you had just finished running the model yourself. In the ArcMap Table of Contents, you see the output layers created by the model (if any). In the Results window, which opens automatically, the model is accessible under the Shared heading. By right-clicking the model, you can choose whether to open it as a tool, edit it in the Model window, or run it immediately with the Re Run choice.

The file-based contents of the package—the model, its toolbox, and all its input and output data—are stored on disk in your C:\Users\<username>\My Documents\ArcGIS\Packages folder (for Windows 7 users).

6) **Back up the model.** You've tested the model and it works. This is a good time to make a backup copy so you can experiment on your own without worrying that you might break something.

Ⓐ In the Catalog window, right-click the *Park Suitability Analysis* model and choose Copy.

Figure 7-86

B Right-click ParkPlanningModels.tbx and choose Paste.

C Rename the copy to Park Suitability Analysis Backup, or any name you like (Figure 7-86).

The name you give the model in the Catalog window is its label, not its actual name. If you check the model properties, you'll see that its name is *ParkSuitabilityAnalysis2*. Models are tools, and tools, to satisfy scripting requirements, don't allow spaces in their names.

D Exit ArcMap. Click Yes if you're prompted to save changes.

You've seen two ways to carry out an analysis project: as a series of stand-alone geoprocessing operations and as a model. Models offer a lot of advantages, but it's an open question as to how fully developed the analysis should be before you turn it into a model.

We recommend planning the analysis in a traditional way: sketching the workflow on paper or a whiteboard, experimenting with geoprocessing tools, and discussing results with colleagues. Wait until you have a pretty firm plan before you take the analysis into ModelBuilder. (And remember that even your "firm plan" will probably change.) If you start with ModelBuilder as your whiteboard, you may find that the model gets too messy in the early planning stages and needs constant corrections. Of course, the more you know about analysis, geoprocessing tools, and ModelBuilder, the sooner you can get your model in good working order. Once that's done, it's relatively easy to add further complexity.

8 Present analysis results

COMMUNICATING YOUR FINDINGS

to a diverse audience of readers with levels of sophistication ranging from public citizens to council members to the City's urban planners who must ultimately design an actual park is no easy task.

Our analysis has yielded a short list of candidate sites that we want to present to the city council and other interested parties. We'll make our presentation in the form of a map that places the results in a meaningful geographic context, addresses the project guidelines, and follows good cartographic design principles.

The deliverable item is an 8½ x 11-inch map that can be printed or viewed on-screen.

But wait, you may be thinking, what's wrong with simply adding the layer of recommended sites to an online basemap, as we already did in Lesson 6? There's nothing wrong with it, and basemaps will be part of our map design, but our map design needs to accomplish a lot of things. We'll be adding inset maps, customizing the positions of labels and other text, adding graphs of statistical data, and incorporating other standard cartographic elements such as scale bars and north arrows.

Our analysis results aren't too complex. We basically just want to show the study area, represented by the LA River buffer zone, and the five proposed park boundaries. We want to set these analytical layers against a backdrop of topography, place names, and major roads. Perhaps the biggest design challenge we face is that these five locations are small parcels of land strung out over a 20-mile stretch of river corridor: how do we map the area of interest without rendering the parks too tiny to see at that scale? The solution lies in the use of inset maps—smaller maps placed within the main map, each one portraying a closeup view of a particular site.

Lesson Eight roadmap

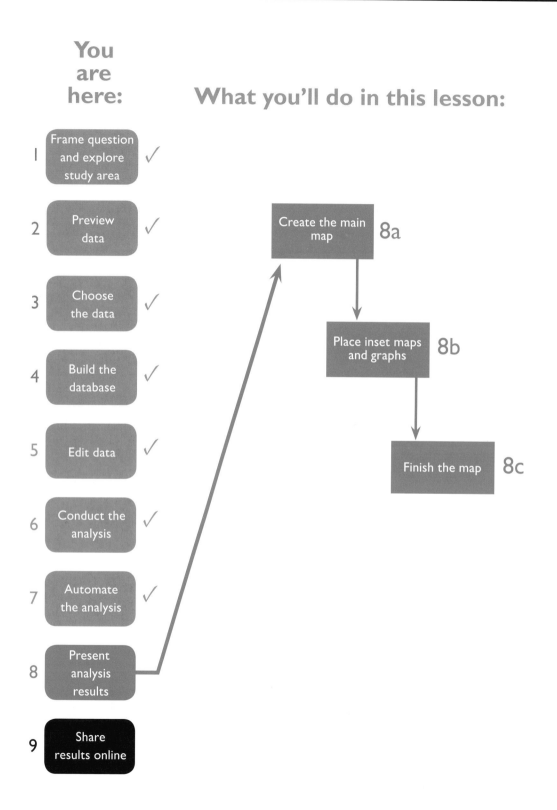

You are here:

1 Frame question and explore study area ✓

2 Preview data ✓

3 Choose the data ✓

4 Build the database ✓

5 Edit data ✓

6 Conduct the analysis ✓

7 Automate the analysis ✓

8 Present analysis results

9 Share results online

What you'll do in this lesson:

Create the main map — 8a

Place inset maps and graphs — 8b

Finish the map — 8c

Another challenge we'll face is how to incorporate comparisons of some of the key demographic variables that drove the selection process: population density, median household income, and the percentage of children.

Cartography involves innumerable choices about color, font, line width, the placement and juxtaposition of elements, and so on. These are often somewhat subjective decisions, with many good (and bad!) possibilities. This lesson takes you through the creation of an exact design (seen in Figure 8-1 at 60% actual size), but along the way you'll have the option of altering things to suit your own taste. We'll try to explain why we made the choices we did, but you don't have to follow the instructions with total fidelity. Experimenting with some of the settings will give you a deeper understanding of how the software works and a chance to make a map that suits your own aesthetic tastes.

Figure 8-1

291

A full presentation of this type in the real world might include a variety of media products, such as large format wall maps, smaller notebook-sized maps, printed reports, digital slide shows, PDF downloads, and web sites. We can't do all these things without writing another book, so we're going to focus on a single map layout that can be printed on a standard letter-sized piece of paper. In Lesson 9, we'll go a step further and take our results to the web, but for now, our working medium will be a sheet of letter-sized paper.

An 8½ by 11-inch map is pretty small, but it would be worse if we were trying to pack a lot more features or a lot of detailed data into this format. We're using this size in recognition of the fact that many people don't have access to larger-format printing devices. And if you can design well in small format, you'll feel that much more comfortable and creative when you begin working in larger formats.

A good way to begin a map layout—just like an analysis—is with a sketch of your basic plan. The sketch for our map is shown in Figure 8-2:

Figure 8-2

Exercise 8a: Create the main map

Beginning with this exercise, we'll be working in ArcMap's layout view. Viewing and working with data is one thing; composing a map for the printed page is something else. In composing maps you must take into account not only your design considerations but also output constraints such as paper size, page orientation, and print margins. Map composition is done in layout view, the graphical, page-layout environment of ArcMap.

1) **Open a map template.** We'll start with a new map document created from a standard template—in this case a landscape (horizontally oriented) page.

- Ⓐ Start ArcMap.
- Ⓑ In the Getting Started dialog box, click Templates in the left-hand window. Under Standard Page Sizes, click North American (ANSI) Page Sizes.
- Ⓒ In the right-hand window, scroll down and click Letter (ANSI A) Landscape to select it (Figure 8-3) and click OK.

Even if your printer uses different sizes, use ANSI A anyway to complete the exercise.

Choosing the template makes ArcMap open in layout view (Figure 8-4).

Select Letter (ANSI A) Landscape

Figure 8-3

rulers

edge of paper

data frame

Layout toolbar

Layers

toggle between data view and layout view

You see an empty data frame on a representation of a piece of paper. A new toolbar, the Layout toolbar, is added to the interface.

Figure 8-4

The idea of a data frame takes on new significance in layout view. In data view (where we've worked until now), a data frame is a more or less abstract "container" of layers. Apart from one or two special properties, like its coordinate system setting that controls on-the-fly projection, we generally don't pay too much attention to it. In layout view, the data frame is visual: it's a rectangular window placed on a virtual sheet of paper and a key element in the map composition.

In the template, the default margins around the data frame are set to one inch on all sides. Most of this space would be wasted in our layout because we're going to place all the map content—data and marginalia both—directly in the data frame. We'll enlarge the data frame, leaving just a ¼-inch margin around the edge of the page.

D From the ArcMap main menu, choose Customize → ArcMap Options.

E In the ArcMap Options dialog box, click the Layout View tab.

F In the Grid area, check the Show box and confirm that the horizontal and vertical spacing values are set to 0.25 inches.

▷ If necessary, choose ¼ in from the drop-down lists.

G In the Snap elements to area, check the Grid box.

H Confirm that your settings match Figure 8-5 and click OK.

Figure 8-5

A pattern of gray dots appears on the virtual page. This is the grid. The outermost dots mark the quarter-inch margin on all sides.

I On the Tools toolbar, click the Select Elements tool ▶, if necessary. Click anywhere inside the data frame to select it.

J Resize the data frame by dragging its blue selection handles.

As you drag the data frame, it snaps to the dots on the grid. Your result should look like Figure 8-6. We don't need the grid anymore, so we'll turn it off.

Ⓚ From the main menu, choose Customize → ArcMap Options.

Ⓛ In the Grid area, uncheck the Show box.

Ⓜ In the Snap elements to area, uncheck the Grid box. Click OK.

Ⓝ Click outside the data frame to unselect it.

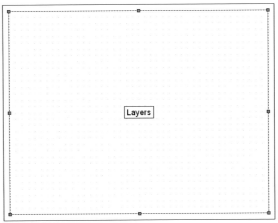
Figure 8-6

2) **Add the LA River to the layout.** The *LARiver* feature class, like all our geodatabase data, is in the State Plane California Zone 5 coordinate system. This is the system we want our map to use. By adding *LARiver* as the first layer, we'll automatically set the data frame to this system and ensure that any basemap layers added thereafter will be projected on the fly to match.

Ⓐ Maximize the ArcMap window.

Ⓑ On the Layout toolbar, click the Zoom Whole Page button .

▷ If the Layout toolbar isn't on the interface, from the ArcMap main menu, choose Customize → Toolbars → Layout.

Ⓒ In the Catalog window, navigate to C:/UGIS/ParkSite/ AnalysisData/ReadyData.gdb.

Ⓓ Drag and drop *LARiver* onto the map.

▷ We want to maximize the layout space, so keep the Catalog window in its Auto Hide (unpinned) state.

Ⓔ In the Table of Contents, open the Symbol Selector for the *LARiver* layer.

Ⓕ In the Symbol Selector, click the predefined River symbol.

Ⓖ Change the width to 4 points and click OK (Figure 8-7).

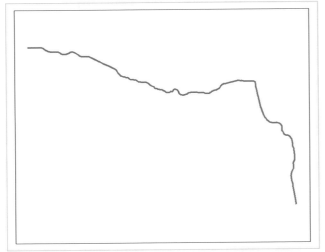
Figure 8-7

This is a good time to make the distinction between the navigation tools on the Layout toolbar (Figure 8-8) and their counterparts on the Tools toolbar.

Figure 8-8

The icons for navigation tools on the Layout toolbar show a "page" in the background.

The layout navigation tools magnify, reduce, and otherwise adjust your view of the virtual page. They do not change the scale or extent of the map within the data frame. Both sets of tools work in layout view, so you need to be careful. (Later on, when we've finalized the map scale and extent, we'll disable the data navigation tools.)

Ⓗ Note the current map scale in the scale box on the Standard toolbar and the current view scale on the Layout toolbar.

Ⓘ On the Layout toolbar, click the Zoom In tool 🔍 and zoom in to the upper left corner of the data frame. It doesn't matter exactly how far.

Notice the effect in Figure 8-9. The rulers have zoomed in, showing you that you're now seeing just a few inches of the page. The line feature also got thicker. The map scale hasn't changed within the data frame, but the view scale is different.

Ⓙ On the Layout toolbar, click the Go Back to Extent button 🔳.

Ⓚ On the Tools toolbar, click the Zoom In tool 🔍 and zoom in on the same upper left corner of the data frame.

▷ You have to begin drawing your zoom rectangle inside the data frame or the tool won't have any effect.

Figure 8-9

This time the map scale changes (you've zoomed in on the data) and the view scale stays the same.

Ⓛ In the Table of Contents, right-click the *LARiver* layer and choose Zoom To Layer.

3) Add a terrain basemap. The Terrain basemap will make an unobtrusive background, with topographic relief that will make the map less starkly flat.

A Add the *Terrain with Labels* basemap (Figure 8-10).

> If you get a Geographic Coordinate Systems Warning, check the "Don't warn me again in this session" box and close the warning.

The basemap is added as two separate layers. The *Basemap* layer, at the bottom of the Table of Contents, represents terrain and ocean. The *Reference* layer, at the top, includes all other features (roads, boundaries, and lakes) as well as labels. Since we plan to use our own reference layers and make our own labels, we don't want this layer in the map.

Figure 8-10

B Remove the *Reference* layer from the Table of Contents.

The labels and features disappear, leaving only the terrain surface and some data credits in the lower-right corner of the map. (We'll work with these service layer credits later in the lesson.)

Before going on, we'll give the map a title and save the map document. This time we'll save the map to the AnalysisData folder instead of the MapsAndMore folder. If we eventually decide to package and share our map (just as we packaged our model in the last lesson), it will be more convenient to have the map and its source data in the same folder.

C From the main menu, choose File → Map Document Properties.

D In the Title box, type **Potential LA River Park Site Locations** (Figure 8-11).

Figure 8-11

E Fill out the remaining map document properties as you did for the Lesson 6 map in Exercise 6d.

The key elements to include are the summary, description, and tags. These are required if you later decide to share the map as a package or publish it as a web service.

F Check the box to store relative paths to data sources.

G Click Make Thumbnail and click OK.

H On the Standard toolbar, click the Save button 💾.

Figure 8-12

❶ In the Save As dialog box, navigate to \UGIS\ParkSite\AnalysisData and save the map document as **Lesson8** (Figure 8-12).

4) Add and symbolize the LA River buffer. In Lesson 6, we created the *LARiverBuffer* feature class. We'll add it now to show the study area and symbolize it with a special gradient fill symbol.

Ⓐ In the Catalog window, expand the AnalysisOutputs geodatabase and add *LARiverBuffer* to the map.

Ⓑ In the Table of Contents, open the Symbol Selector for the *LARiverBuffer* layer.

Ⓒ Click Edit Symbol to open the Symbol Property Editor.

We opened the Symbol Property Editor once before, in Lesson 1, to add a halo to a label. That earlier effect and the gradient fill we're about to use now barely scratch the surface of what you can do with symbols. See the topic *Editing symbol properties* for more information.

Right-click to toggle off Graphic View.

Figure 8-13

Select this color ramp (Green Bright).

Ⓓ In the Symbol Property Editor, click the Type drop-down arrow and choose Gradient Fill Symbol.

Ⓔ On the Gradient Fill tab, click the Style drop-down arrow and choose Buffered.

Ⓕ Set the Intervals to **40**.

Ⓖ Set the Percentage to **20**.

Ⓗ Click the Color Ramp drop-down arrow and choose Green Bright (Figure 8-13).

▷ Or right-click the color ramp, uncheck Graphic View, and choose the ramp by name.

❶ Click OK to close the Symbol Property Editor.

Ⓙ In the Symbol Selector, set the outline color to No Color.

Ⓚ Click OK to close the Symbol Selector.

A gradient fill changes color or intensity across the symbol. Of the four styles, we chose Buffered because it makes the pattern radiate outward along the length of the river. The number of intervals determines the smoothness of the effect: with more intervals, color transitions are smoother, but the symbol takes longer to draw. The percentage controls the pattern of color saturation. You may want to experiment with different settings or ramps. The preview in the upper left corner of the Symbol Property Editor helps you anticipate the effects of your settings.

L Open the layer properties for the *LARiverBuffer* layer and click the Display tab.

M In the Transparent % box, type **30** and click OK.

The result is a smooth gradient fill with terrain showing through around the outer edges of the buffer (Figure 8-14). It looks good, but it also has a subtle analytic effect. Our analysis prefers candidates closer to the river, and the symbol is a visual reminder that a candidate in a dark green part of the zone is closer to the river than a candidate in a light green part.

Why make the buffer green? One reason is that we have many layers to symbolize and green works well in combination with the other colors we're going to use. On a psychological level, we're also suggesting that this area is verdant, riparian habitat—whereas, in reality, much of it is either densely developed residential/commercial land or industrial blight. Are we misleading our audience or just being optimistic? Whichever way you think of it, you need to be aware as a cartographer that symbols may have hidden meanings as well as overt ones.

Figure 8-14

5) **Set basemap transparency.** In preparation for the next step, we'll make the terrain basemap partially transparent.

A Open the layer properties for the *Basemap* layer and click the Display tab.

B In the Transparency % box, type **40** (Figure 8-15) and click OK.

On the map, the only difference you might notice is that the *Basemap* layer is a little bit lighter. The transparency effect won't become obvious until you add some data underneath it.

Figure 8-15

6) **Add and symbolize the LA city limits.** We only considered locations within Los Angeles itself, so it will be helpful to show the city limits on the map.

A In the Table of Contents, collapse the *Basemap* layer.

That way you won't accidentally add the *Cities* layer as a sublayer of the basemap.

B In the Catalog window, under ReadyData.gdb, drag *Cities* to the Table of Contents and drop it at the bottom, underneath the *Basemap* layer (a horizontal black bar marks the spot).

When you drop a layer on the map, it takes its "natural" place in the top-to-bottom order (point, line, polygon, and raster). When you drop it in the Table of Contents, it goes exactly where you put it. It may seem counterintuitive to place a layer of features below a basemap, but doing so will allow us to create a nice, unobtrusive effect.

C Open the layer properties for the *Cities* layer and click the Definition Query tab.

D Click Query Builder.

E In the Query Builder dialog box, create the following expression:

`"NAME" = 'Los Angeles'`

This is the very same definition query we created at the beginning of Lesson 1.

F Check your query against Figure 8-16 and click OK on the Query Builder dialog box.

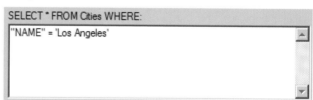

SELECT * FROM Cities WHERE:

"NAME" = 'Los Angeles'

Figure 8-16

G On the Layer Properties dialog box, click Apply and click the Symbology tab.

H On the Symbology tab, click the symbol patch to open the Symbol Selector.

I Set the fill color to Yucca Yellow (Figure 8-17) and the outline color to No Color.

J Click OK on both the Symbol Selector and the Layer Properties dialog boxes.

The result is a subdued yellow filtered through the semi-transparent basemap. White space shows through areas that aren't part of Los Angeles proper. Some of these areas are pretty famous places, such as Beverly Hills and Santa Monica, and we'll label them in Exercise 8c.

7) Add and symbolize the LA County boundary. The white showing through is a bit stark, so we'll mute it by putting the *Counties* layer underneath *Cities*.

A In the Catalog window, under ReadyData.gdb, drag *Counties* to the bottom of the Table of Contents, underneath the *Cities* layer.

B Open the Symbol Selector for the *Counties* layer.

C Set the outline color to No Color.

D For the fill color, at the bottom of the color palette, click More Colors to open the Color Selector.

Figure 8-17

Editing symbol properties

When you add a layer to ArcMap, it's symbolized with a simple symbol type. Point layers are symbolized with simple marker symbols, polylines with simple line symbols, and polygons with simple fill symbols. Simple symbol types have just a few editable properties, such as size, width, and color, which can be changed on the Symbol Selector. To make more complex symbols, open the Symbol Property Editor by clicking Edit Symbol on the Symbol Selector, and change the symbol type at the top of the dialog box. (The predefined symbols that you choose in the Symbol Selector are often nonsimple types.)

Symbol types for point layers include character, picture, and arrow markers. A character marker can be any glyph from any font in your system font folder. A picture marker can be an image in any supported file format (.bmp, .emf, .png, .jpg, .jpeg, or .gif). In this example, schools are symbolized with a character marker symbol. The font has been set to ESRI Default Marker, one of many symbol fonts installed with ArcGIS for Desktop. The symbol is the glyph circled in blue.

To look for predefined symbols, such as this one for schools, type a keyword in the search box at the top of the Symbol Selector.

Symbol types for polygon layers include line, marker, picture, and gradient fills. A line fill symbol fills polygons with lines of any width, spacing, angle, and color. You can also fill polygons with markers or pictures. A gradient fill symbol fills polygons with a color ramp. Here, a line fill symbol is used for a buffer zone along the LA River. The symbol is built from two superimposed graphic "layers."

The line pattern draws on top of a solid color.

Symbol types for polyline layers include cartographic, hash, marker, and picture. Cartographic line symbols let you make dash patterns and other effects. Marker and picture line symbols let you place recognizable icons at intervals along a line (to symbolize a bus or bike route, for example). Here, a cartographic line symbol is used to make a dash pattern for intermittent streams.

Figure 8-18

Type in values

preview of
your color

... or move
the sliders.

In the Color Selector, you can mix your own colors using either the RGB or CMYK color models. We'll work with RGB, which is the model used for computer monitors. (The HSV model, which is similar to RGB, is also available.) When you design a map for printed output, you have to expect that there will be color shifts. Since these shifts vary according to the output device or printing process, you should make test prints of your map as it develops and modify the on-screen colors as needed.

ⓔ In the Color Selector, make sure the Color tab is selected and the drop-down list is set to RGB.

ⓕ Type the following values: R = **235**, G = **235**, B = **250** (Figure 8-18). Click OK on the Color Selector and the Symbol Selector.

On the map, you should see a subtle contrast between the City of Los Angeles, the other land area, and the water of the Pacific Ocean (Figure 8-19).

Figure 8-19

8) Add and symbolize major roads. Road networks strongly influence our sense of local geography and are important for orientation on many maps.

ⓐ In the Catalog window, under ReadyData.gdb, add *MajorRoads* to the map.

 ▷ It should appear at the top of the Table of Contents. If it doesn't, move it there now.

ⓑ Open the attribute table for the *MajorRoads* layer and scroll across the attributes.

The table contains the name of each road and various other descriptors.

ⓒ Right-click the ROAD_TYPE field heading and choose Sort Ascending (Figure 8-20).

ⓓ Scroll down through the table and note the values in this field.

There are three road type values. "Highway" is for secondary state and county highways; "Limited Access" is for big freeways and interstate highways; and "Road" typically designates important streets. We'll symbolize the layer on the values in this field, using different line thicknesses.

Take note also of the HWY_SYMBOL attribute next to ROAD_TYPE. This field stores numeric designators, primarily for the freeways and highways. We'll use this field to label the roads in Exercise 8c.

ⓔ Close the attribute table.

ⓕ Open the layer properties for the *MajorRoads* layer. Click the Symbology tab, if necessary.

ⓖ In the Show box, click Categories and accept the default Unique Values option.

ⓗ Set the Value Field to ROAD_TYPE.

ⓘ Click Add All Values (Figure 8-21).

ⓙ In the Symbol column, right-click the Highway symbol and choose Properties for Selected Symbols to open the Symbol Selector.

ⓚ Set its color to Gray 20% and its width to 2. Click OK.

ⓛ For the Limited Access symbol, set the color to Gray 20% and the width to 4.

ⓜ For the Road symbol, set the color to Gray 20% and the width to **0.5**.

ⓝ In the Value column, click Limited Access to highlight it and click the Up arrow on the right side of the dialog box (Figure 8-22).

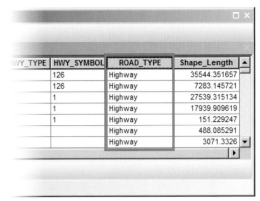

Figure 8-20

The Count column shows the total number of features in the layer and how many features have each value.

Figure 8-21

Figure 8-22

Figure 8-23

Uncheck

In the Table of Contents, the values will now be listed in descending order of thickness, which looks good. Manipulating this list order, however, has no effect on the order in which the features are drawn. In other words, it does not make Limited Access features draw on top of Highway features.

O In the Symbol column, uncheck the box for <all other values> (Figure 8-23).

Since there aren't any other values (the Count is 0), the map will look the same whether or not the box is checked. The point of unchecking the box is to prevent the <all other values> label from showing up in the Table of Contents.

Symbol levels

Sometimes you want to control the drawing order of features within a layer. For example, suppose you have a *Roads* layer that includes paved roads, unpaved roads, and foot trails. If these values are symbolized with distinct colors or dash patterns, it will be obvious, wherever features of different types cross, that one is on top and one is underneath. You might want paved roads to draw on top of unpaved roads and unpaved roads to draw on top of trails. You can control this feature drawing hierarchy within a layer by setting symbol levels. On the Symbology tab, click the Advanced button to access the Symbol Levels dialog box. (We avoided this problem with the *MajorRoads* layer because all three road types have the same symbol color.)

P Click OK to close the Layer Properties dialog box.

The *MajorRoads* layer adds context to the map without overwhelming it (Figure 8-24) and the symbology is fairly self-explanatory.

Figure 8-24

9) Set the map extent. At the moment, the map is zoomed to the extent of the *LARiverBuffer* layer. While this is our area of interest, we need to make a little more room around the edges for inset maps and graphs.

Ⓐ On the Standard toolbar, set the map scale to **1:160,000**.

At this scale, one inch on the map equals about two and a half miles on the ground. With your ArcMap window maximized, the map extent should be close to Figure 8-25, although the result depends on things like the size and screen resolution of your monitor. For our purposes, the extent is more important than the scale.

Figure 8-25

Ⓑ If necessary, use navigation tools on the Tools toolbar (not the Layout toolbar!) to approximate the map extent in the figure.

Now we'll lock this extent so we don't change it by mistake.

Ⓒ Open the data frame properties and click the Data Frame tab.
  ▷ How? In the Table of Contents, right-click the name of the data frame (Layers) and choose Properties.
Ⓓ On the Data Frame tab, click the Extent drop-down arrow and choose Fixed Extent (Figure 8-26).
Ⓔ Click the Specify Extent button (Figure 8-27).

Figure 8-26

Figure 8-27

The default setting, Current Visible Extent, is what we want.

Ⓕ Close the Data Frame - Fixed Extent dialog box.
Ⓖ Click Apply on the Data Frame Properties dialog box, but leave it open.
  ▷ If you get a coordinate system warning, check the "Don't warn me again in this session" box and click Yes.

❶ Click the General tab and change the name of the data frame from Layers to **Main Map**. Click OK on the Data Frame Properties dialog box.

In the Table of Contents, the data frame is renamed. On the Tools toolbar, the navigation tools are disabled because you fixed the extent. Now you can't zoom or pan the map.

10) Add and symbolize recommended sites. The five recommended sites are the most important features on our map, but they're so small and spread out that we need to do some creative symbology to make them show up. In our sketch at the beginning of the lesson, we used red circles to represent the sites. We'll create that effect here by symbolizing the *RecommendedSites* layer (a polygon layer) with graduated symbols. We used this technique before in Exercise 1b to show income values within block groups as circles of varying size. Our purpose here is more unorthodox. Because we can easily enlarge the symbols to a size that will draw the map reader's attention, we'll use them as substitutes for the small site polygons. It will still be important to show the actual shape and size of the sites, and to do that, we'll create inset maps in the next exercise.

Figure 8-28

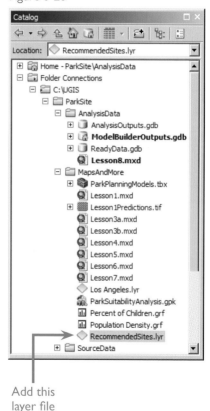

Add this
layer file

❶ In the Catalog window, expand Folder Connections, if necessary, and navigate to the MapsAndMore folder.

❶ Add the layer file *RecommendedSites.lyr* to the map (Figure 8-28).

▷ If you don't have this file, you can download it from the Lesson 6 results at the Understanding GIS web site.

In the Table of Contents, the layer appears directly above the *LARiverBuffer* layer. It has the bright green outline and hollow fill we gave it in Lesson 6.

❶ Open the attribute table for the *RecommendedSites* layer.

❶ Confirm that it shows the field aliases and formatting from Lesson 6 and close the table.

The point of using the layer file is not for its symbology (which we're about to change) but for its attribute formatting. We'll take advantage of this in the next exercise when we work with graphs and inset maps.

❶ Open the layer properties for *RecommendedSites* and click the Symbology tab, if necessary.

❶ In the Show box, click Quantities. Under Quantities, click Graduated Symbols.

❶ In the Fields area, click the Value drop-down list and choose SITEID.

Each of the five sites is assigned a symbol. By default, the symbols increase in size from 4 to 18 points. That would be fine if we were representing a range of attribute values, but it doesn't suit our special case. We want all our "graduated" symbols to be the same size.

Figure 8-29

H Click the Template button to open the Symbol Selector.

I In the list of symbols, click Circle 1.

J Make the color Mars Red (Figure 8-29) and click OK on the Symbol Selector.

K Click the Background button to open the Symbol Selector again.

L Change the fill color to No Color. Change the outline color to No Color as well and click OK on the Symbol Selector.

M On the Symbology tab, in the Symbol Size boxes, replace both the "from" and "to" values with **22**.

N Check your settings against Figure 8-30 and click OK.

Figure 8-30

The sites will be symbolized with red circles at the center of each polygon. Because the background is transparent, only the circles will be visible.

On the map, the symbols stand out prominently (Figure 8-31). Cartographers often use very bright colors (called spot colors) for the features they most want to call attention to on the map.

Figure 8-31

11) Label the sites. We'll use the SITEID field again to label the sites.

Ⓐ Open the layer properties for *RecommendedSites* and click the Labels tab.

Ⓑ In the upper left corner of the dialog box, check the "Label features in this layer" box.

Ⓒ Click the Label Field drop-down arrow and choose SITEID.

Ⓓ In the Text Symbol area, accept the default font (Arial), but change the size to 9 and click the Bold button **B** .

Ⓔ Set the color to Arctic White.

Ⓕ Check your settings against Figure 8-32 and click OK.

Figure 8-32

On the map, the symbols are labeled with their SITEID numbers, but there's a slight problem. Reading from right to left, the red circles are not labeled sequentially. Instead of "1, 2, 3, 4, 5" they read "1, 2, 4, 3, 5" as Figure 8-33 shows. This just happens to be the outcome of our geoprocessing operations in Lesson 6, but it's displeasing to our sense of order.

12) Edit the *RecommendedSites* attribute table. We'll edit the table to renumber Sites 3 and 4.

Ⓐ In the Table of Contents, right-click the *RecommendedSites* layer and choose Edit Features → Start Editing.

This is another way to start an edit session. Note that the Editor toolbar has been automatically added to the interface.

Figure 8-33

B Open the *RecommendedSites* attribute table (Figure 8-34).

C In the SITEID field, click the cell containing the value "3." Highlight and delete the value and type **4** in its place.

D Press Enter to select the next cell down. Type **3** and press Enter. Confirm that your table looks as shown in Figure 8-35.

E On the Editor toolbar, click the Editor menu and choose Save Edits. Click the Editor menu again and choose Stop Editing.

On the map, the sites are now labeled in the desired order.

F Close the table window.

▷ Optionally, remove the Editor toolbar. (You won't need it again.)

G In the Table of Contents, collapse the *MajorRoads* and *RecommendedSites* layers so they don't take up so much room.

H Save the map.

I If you're continuing to the next exercise now, leave ArcMap open; otherwise, exit ArcMap.

In this lesson, we've symbolized the layers that make up the main map. In the next exercise, we'll add inset maps to show close-ups of the sites and graphs that compare the neighborhood demographics.

Exercise 8b: Place inset maps and graphs

An inset map is a small, supplementary map that shows part of the main map's area at a larger scale. It reveals detail not apparent at the scale of the main map. (An overview map, by contrast, shows the main map at a smaller scale to place it in the context of its surroundings.)

We want map readers to have a close up look at each of the sites, so we'll create five large-scale inset maps. We can't simultaneously set different map scales within a single data frame (except with viewer windows), so we need to insert some new data frames. In fact, we'll insert five of them and place them around the main map as shown in our layout sketch. After we finish the inset maps, we'll add bar graphs comparing median income, the percentage of children, and population density for the sites.

1) **Insert a data frame.** We'll insert a new data frame to contain our first inset map.

A If necessary, start ArcMap and open Lesson8.mxd from the list of recent map documents.

B From the main menu, choose Insert → Data Frame.

A new data frame is placed in the center of the map display (Figure 8-36) and selected. In the Table of Contents, its name is boldfaced to show that it's the active data frame.

Figure 8-34

These two values need to be switched

Figure 8-35

Figure 8-36

Figure 8-37

ⓒ Open the data frame properties for New Data Frame.

ⓓ In the Data Frame Properties dialog box, click the Size and Position tab.

ⓔ In the Size area, set the Width and Height to **1.5**.

The values default to inches (in).

ⓕ In the Element Name box, rename the data frame to **Site 1**.

Don't worry about the X and Y position values: we'll position the data frame manually.

ⓖ Check your settings against Figure 8-37 and click the Frame tab.

ⓗ Click the Border drop-down arrow and click the 0.5 Point border (Figure 8-38).

ⓘ Click the Color drop-down arrow and set the color to Gray 40%. Click OK.

On the map, the data frame is resized and renamed.

2) Add the recommended sites. We'll add the *RecommendedSites* layer to each new data frame; then zoom in on a different site in each data frame.

ⓐ In the Catalog window, under MapsAndMore, drag and drop *RecommendedSites.lyr* onto the layout.

It doesn't matter where you drop it—the layer will be added to the active Site 1 data frame. The layer is listed in the Table of Contents, but on the map the features are barely visible in the transparent data frame. (The data frame label goes away now that the data frame has content.)

ⓑ Open the attribute table for the *RecommendedSites* layer in the Site 1 data frame.

  ▷ Make sure you're not opening it from the Main Map data frame.

ⓒ If necessary, move the table away from the Site 1 data frame.

ⓓ In the attribute table, right-click the small gray box next to the first record (SITEID 1) and choose Zoom To (Figure 8-39).

Figure 8-38

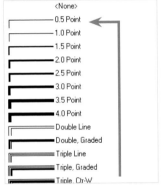

Choose the 0.5 point border.

Right-click here and choose Zoom To.

On the map, the feature fills the data frame (Figure 8-40).

ⓔ Close the attribute table.

Figure 8-39

Figure 8-40

3) Add basemap imagery to the inset map. At the large scale of the inset map, imagery will show the site as it really looks. (Keep in mind that basemap layers are updated periodically, so your imagery may look different.)

Ⓐ Add the *Imagery* basemap to the Site 1 data frame (Figure 8-41).

When you add a basemap layer, the data providers are credited on the basemap itself. In data view, these credits appear as an icon ⊟ in the lower right corner of the data frame. (Clicking the icon opens the credits in a window.) In layout view, the credits are displayed as text. You may have noticed them before in the main map, but you can't miss them in the inset map, which is overwhelmed by the text. These credits can't be deleted, but they can be manipulated. In Exercise 8c, we'll incorporate them into some other data acknowledgments, but for the time being, we need to move them out of the way.

Ⓑ On the main menu, choose Insert ➤ Dynamic Text ➤ Service Layer Credits.

Both sets of credits are moved out of their respective data frames and into a text box on the layout (Figure 8-42). The text box is selected.

Figure 8-41

Figure 8-42

Ⓒ Drag the text box anywhere outside the boundary of the layout page (Figure 8-43).

Ⓓ Click some empty white space to unselect the box; then click the Zoom Whole Page button ⬚ on the Layout toolbar.

With the service layer credits out of the way, we can turn our attention back to the inset map. The bright green outline for the park site looks pretty good, but yellow might look even better.

Drag and place outside of the page edge.

Service Layer Credits: Sources: USGS, Esri, TANA, AND
Source: Esri, DigitalGlobe, GeoEye, i-cubed, USDA, USGS, AEX, Getmapping, Aerogrid, IGN, IGP, swisstopo, and the GIS User

Figure 8-43

Figure 8-44

ⓔ In the Table of Contents, under the Site 1 data frame, open the Symbol Selector for *RecommendedSites*.

ⓕ Change the Outline Width to 3 and the Outline Color to Solar Yellow (Figure 8-44). Click OK.

Your inset map should look like Figure 8-45.

Figure 8-45

4) **Label the site.** We want to label the recommended sites with their sizes so the map reader knows how big they are. As we get closer to finishing the map, we'll be increasingly concerned with the exact placement and appearance of text, and for that reason we'll switch from the default Standard Label Engine to the Maplex Label Engine. Eventually, we'll want to take complete manual control of at least some of our map text, and we'll see how to do that in Exercise 8c.

We previously used the Maplex Label Engine in Exercise 1b to stack park labels automatically. The Maplex Label Engine has many other advanced properties for placing labels inside and near features, for making them fit in tight spaces, for duplicating them judiciously in large, long, or oddly shaped features, and for resolving conflicts with other labels. We'll make use of some of these capabilities as we label the sites, and, later on, the cities and major roads.

ⓐ Open the data frame properties for Site 1 and click the General tab.

Figure 8-46

ⓑ Near the bottom of the dialog box, click the Label Engine drop-down arrow and choose Maplex Label Engine (Figure 8-46). Click OK.

ⓒ Open the layer properties for *RecommendedSites* and click the Labels tab if necessary.

D Check the Label features in this layer box.

E In the Text Symbol area, leave the font set to Arial but change the size to 9.

F Set the color to Solar Yellow and click the Bold button **B**.

G Next to the Label Field setting, click Expression.

A label expression lets you add custom text to an attribute value to make a more informative label.

H In the Label Expression dialog box, highlight and delete the current expression.

I In the list of fields, double-click ACRES to add it to the expression box.

J Type the rest of the expression as follows: **& " Acres"**.

Check your expression against Figure 8-47.

The expression tells ArcMap to take the value from the ACRES field and tack on a space with the word "Acres." Without the expression, the map reader wouldn't know what the units were.

Figure 8-47

K Click OK on the Label Expression dialog box and the Layer Properties dialog box.

The labeled feature should look similar to Figure 8-48.

Since there isn't enough room to put all the inset maps right next to their sites, we'll move them to convenient locations, as shown in the layout sketch at the beginning of the lesson. Later, we'll connect them to the sites with leader lines.

L On the map, select the Site 1 data frame. (You should see the blue selection handles.)

M Drag the data frame to the lower right corner of the main map as shown in Figure 8-49.

Figure 8-48

Service Layer Credits:
TANA, AND
Source: Esri, DigitalGlo
USDA, USGS, AEX, Ge
IGN, IGP, swisstopo, a

Figure 8-49

Figure 8-50

Copy the Site 1 data frame and change its name to Site 2.

Figure 8-51

5) Create the rest of the inset maps. The rest of the inset maps will be the same size as Site 1 and will have the same layers with the same symbology. Instead of inserting new data frames, we'll just copy and paste the one we have and zoom it to the appropriate site.

Ⓐ On the map, right-click the Site 1 data frame and choose Copy from the context menu.

Ⓑ Right-click anywhere on the main map and choose Paste.

A new data frame, identical to Site 1 (and with the same name), appears on the layout and in the Table of Contents.

Ⓒ In the Table of Contents, click on the name of the active Site 1 data frame (the new, boldfaced one) to highlight it. Click again to make its name editable.

Ⓓ Rename the data frame to **Site 2** and press Enter (Figure 8-50).

Ⓔ On the layout, drag the Site 2 data frame to the approximate position shown in Figure 8-51.

The Site 2 inset map is still zoomed in on the Site 1 feature.

Ⓕ In the Table of Contents, in the Site 2 data frame, open the attribute table for *RecommendedSites*.

In the last exercise, we edited the Site ID values of the third and fourth records in the table. It will be easier to zoom to the right feature in each inset map if we sort the SITEID field.

Ⓖ Right-click the SITEID field heading and choose Sort Ascending.

In the table, the Site ID values are now listed in order (Figure 8-52). (If you don't want to be distracted by the other ID numbers, you can right-click the OBJECTID field heading and choose Turn Field Off. You can turn it on again later on the Fields tab of the Layer Properties dialog box.)

OBJECTID *	Shape *	SiteID	Acres
1	Polygon	1	5.7
2	Polygon	2	1.7
4	Polygon	3	3.4
3	Polygon	4	2
5	Polygon	5	1.7

Figure 8-52

Ⓗ In the attribute table, right-click the small gray box next to the second record (SITEID 2) and choose Zoom To.

The inset map zooms in to Site 2 (Figure 8-53). In each inset map, we're letting the site boundary fill the data frame. That means the inset maps will be at different scales from one another—a fact we'll note on the map later.

I Close the attribute table.

J Repeat the process to create three more inset maps. Use Figure 8-54 to guide your placement of the data frames. In summary, the steps are:

- Copy and paste the data frame you just made.
- Rename the new data frame.
- Drag it to the appropriate place on the layout.
- Open the *Recommended Sites* attribute table and zoom to the feature with the matching SITEID number.

Figure 8-53

Figure 8-54

K When you're finished, save the map.

L If necessary, in the Table of Contents, drag and drop the inset map data frames into sequential order.

▷ Optionally, collapse them in the Table of Contents.

6) **Set label placement properties.** Four of the site labels are well-placed, but the one at Site 5 is squeezed against the edges of the polygon boundary. It will look better if we turn it sideways.

A If necessary, make Site 5 the active data frame.

▷ How? Right-click the data frame name in the Table of Contents and choose Activate.

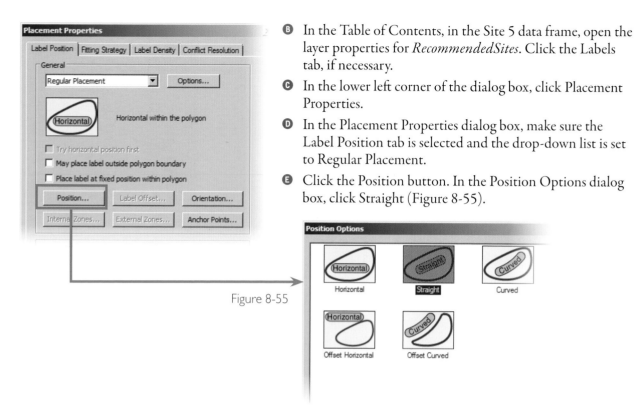

Figure 8-55

B In the Table of Contents, in the Site 5 data frame, open the layer properties for *RecommendedSites*. Click the Labels tab, if necessary.

C In the lower left corner of the dialog box, click Placement Properties.

D In the Placement Properties dialog box, make sure the Label Position tab is selected and the drop-down list is set to Regular Placement.

E Click the Position button. In the Position Options dialog box, click Straight (Figure 8-55).

Figure 8-56

The straight setting orients the label to the longest axis in the polygon feature.

F Click OK on all open dialog boxes.

The label for Site 5 is rotated to fit the feature better (Figure 8-56). Depending on your system and settings, some of your labels may look slightly different (for example, the text may not always be stacked).

7) Create a graph of median household income. We know each site meets our minimum demographic requirements, but it would be nice to compare them graphically to find out if one or more stand out as especially well-qualified. In this step, we'll create a bar graph of median household income. Later, we'll add preexisting graphs to show population density and the percentage of children.

A Activate the Main Map data frame.

B From the main menu, choose View → Graphs → Create Graph to open the Create Graph Wizard.

C At the top of the dialog box, click the Graph type drop-down arrow. In the Bar folder, choose Horizontal Bar.

D Click the Layer/Table drop-down arrow and choose *RecommendedSites*.

E Set the Value field to Median Household Income.

ⓕ Set the Y field to Site ID.

ⓖ Leave the Y label field set to <None>.

In the preview window, the graph is starting to take shape. The five sites are listed on the y-axis and incomes are shown on the x-axis.

ⓗ Uncheck the "Add to legend" box.

ⓘ Click the Color drop-down arrow and choose Custom.

ⓙ Click the color drop-down arrow that appears next to this setting. On the color palette, choose Cretan Blue (Figure 8-57) or another color you think will look good on the map.

ⓚ Compare your graph preview and settings to Figure 8-58 and click Next.

Figure 8-57

Figure 8-58

ⓛ On the next panel, in the General graph properties, replace the default title (Graph of RecommendedSites) with **Median Household Income** (Figure 8-59).

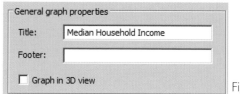

Figure 8-59

The title makes the x-axis label redundant. We'll address this shortly.

ⓜ Click Finish.

Figure 8-60

The graph is displayed in a floating window that is not yet part of the map layout (Figure 8-60). You can move it outside the virtual page and even outside the ArcMap window (except that ArcMap should be maximized). The graph is also live-linked to the data: if you were to edit a value in the attribute table, the graph would update automatically.

8) Set advanced properties for the graph. The three graphs are going to be fairly small design elements in the layout. We'll make the axis labels a little bigger so they'll be legible on the final map.

Ⓐ Right-click anywhere on the graph window and choose Advanced Properties to open the Editing dialog box.

Ⓑ In the left pane of the Editing dialog box, under Chart, expand Axis. Click Left Axis to select it.

Figure 8-61

On the right side of the dialog box, a set of tabs appears with all the possible settings for the left axis (that is, the y-axis).

Ⓒ Click the Labels tab (top row) and the Text tab (next row).

Ⓓ If necessary, click the Font tab (bottom row). Change the Size to 10 and the Style to Bold. (Figure 8-61).

This makes the SiteID numbers on the left axis more prominent.

Ⓔ In the left-hand pane, click Bottom Axis.

Ⓕ Click the Title tab (top row) and click the Format tab underneath it.

Ⓖ Click the Text tab (third row) and make sure the Font tab (fourth row) is selected.

Ⓗ Change the Size to 10 and check the Bold box.

This makes the bottom axis title more prominent. As noted, however, this axis title is redundant.

Ⓘ Click the Style tab (second row).

Ⓙ Replace the title (Median Household Income) with **Dollars ($)** (Figure 8-62).

Ⓚ On the Editing dialog box, click Close to accept your changes.

Figure 8-62

Now that the graph looks the way we want, we can make it an element of the layout.

ⓛ Right-click anywhere on the graph and choose Add to Layout from the context menu.

The graph window remains open and the graph also appears as a selected element on the layout (Figure 8-63).

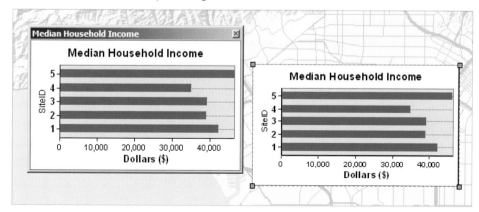

Figure 8-63

The graph is saved with the map document and can be accessed at any time from the View menu. (It can also be saved as a file to disk.) The layout element can now be moved, resized, and otherwise treated as a graphical object on the virtual page. It still maintains its connection to the graph. If you edit properties in the graph window, or edit data values in the *RecommendedSites* attribute table, these changes will be reflected in the layout element.

ⓜ Close the Median Household Income graph window.

On the layout, the graph is too big.

ⓝ Right-click the layout graph and choose Properties. On the Properties dialog box, click the Size and Position tab.

ⓞ In the Size area, change the current Width to **1.75** in and make sure the Preserve Aspect Ratio box is checked. Click Apply.

ⓟ Check your settings against Figure 8-64 and click OK.

Changing the width also changes the height when Preserve Aspect Ratio is checked.

Figure 8-64

For now, you can just leave the layout graph where it is on the page. We'll add two more graphs in the next step and then move and position all the graphs together.

9)   Add graphs of population density and percentage of children. The two remaining graphs have already been created and we'll just add them to the layout. If you installed your project data to the recommended directory (C:\UGIS), this will go smoothly and you can begin the lettered steps now.

If you installed the data to a different location, you'll get a warning message when you try to load the graphs. That's because the graphs are following a hard-coded path (C:\UGIS\Parksite\AnalysisData\AnalysisOutputs.gdb\RecommendedSites) which, in your case, doesn't lead to the data. You have two options. One is to click No on the warning and create the graphs yourself. Follow the same steps you used to create the median household income graph, making appropriate changes to the value field, graph title, and bottom axis title.

Alternatively, click Yes on the warning. This will add the graph window to ArcMap with the correct title and axis labels but no data. It will also add a *RecommendedSites* layer to the top of the Table of Contents with a red exclamation point (which means the layer can't find its data source). You then proceed as follows:

- Open the layer properties for this new *RecommendedSites* layer.
- Click the Source tab. Click Set Data Source. Navigate to the location of your *RecommendedSites* feature class, select it, and click Add. Click OK.
- Right-click the graph and choose Properties. Confirm or make these settings: the Layer/Table is *RecommendedSites*, the Value field is PCT UNDER 18, and the Y field is SiteID.
- Click OK.

The graph should update correctly. You can then remove the extra *RecommendedSites* layer (the new one that was added when you loaded the graph) from the Table of Contents.

Ⓐ From the main menu, choose View → Graphs → Load Graph.

Ⓑ In the Open dialog box, navigate to C:\UGIS\ParkSite\ MapsAndMore.

  ▷ You may want to start by clicking the Look in drop-down arrow at the top of the dialog box.

Ⓒ Click *Percent of Children.grf* to select it; then click Open to add the graph window to ArcMap (Figure 8-65).

Ⓓ Right-click the Percent of Children graph and choose Add to Layout; then close the graph window.

Figure 8-65

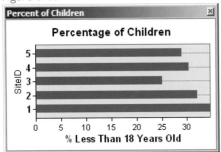

ⓔ In the same way, load *Population Density.grf* (Figure 8-66).

ⓕ Add the Population Density graph to the layout and close the graph window.

The layout graphs may all be on top of each other.

ⓖ Drag the Population Density and Percentage of Children graphs to other parts of the layout so you can see all three graphs.

ⓗ Click the Population Density graph to select it. Hold down the Shift key and click first on the Percentage of Children graph, and then the Median Household Income graph.

All three graphs should be selected (Figure 8-67).

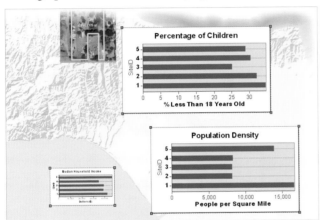

Figure 8-67

ⓘ Make sure that Median Household Income is the last selected element. (It should have blue selection handles; the other two graphs should have green handles.)

▷ The selection order affects the next step, so, if necessary, unselect the graphs and select them again.

ⓙ Right-click any of the selected elements and choose Distribute → Make Same Size.

The two big graphs are resized to match the little (last selected) one.

ⓚ Click outside the layout page to unselect the elements.

Managing graphs

Graphs windows in a map document can be opened at any time from the Graphs command on the View menu. Open the Graph Manager (View → Graphs → Manage Graphs) to list all graphs and access their properties. Graphs can be saved as files with the extension .grf and loaded into other ArcMap documents. Graphs can also be exported as images in standard file formats. A graph references a layer or table in the map document in which it is created. That layer or table in turn references a dataset on disk. When a graph created in one map is saved and loaded into another, the second map needs to have the same layer, with the same value field, on which the graph was originally based.

10) Place the graphs. In the layout sketch, the graphs are arranged vertically on the left side of the layout. We'll align them just a tiny bit away from the edge of the layout using a guide.

Ⓐ At the top of the layout page, hover over the top ruler and click to create a guide.

Guides are nonprinting lines, light blue on the layout, that you can use to align and position elements. The current position of the guide is marked on the top ruler with a small gray arrow.

Ⓑ On the top ruler, click and drag the guide to the 0.30 inches mark.

As you drag the guide, its current position on the ruler is shown as a tip (Figure 8-68).

A tip shows the current position.

Guides are shown as thin blue lines. When being moved, they change to a dotted black line.

Figure 8-68

Figure 8-69

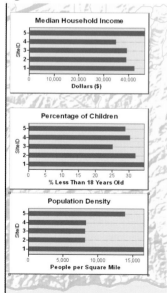

It doesn't matter what the top-to-bottom order of the graphs is.

Ⓒ If necessary, click the Select Elements tool ▸ on the Tools toolbar. Click on the Site 5 inset map in the upper left corner of the layout to select it.

Ⓓ Drag the inset map away from the guide and then back so that it snaps to the guide.

Ⓔ Click and drag the three graphs, one at a time, until they snap to the guide.

Ⓕ Arrange them more or less as shown in Figure 8-69.

Ⓖ Hold the Shift key and select all three graphs (in any order).

Ⓗ Right-click any selected graph and choose Distribute → Distribute Vertically.

The vertical spacing between the graphs is made equal.

Ⓘ Right-click again on a graph and choose Group.

Ⓙ Click outside the layout page to unselect the grouped elements.

In the next exercise, we'll label some features, add a title, descriptive text, and some common map elements such as a scale bar and north arrow. At this point, the layout should look like Figure 8-70.

Figure 8-70

K Save the map.

L If you're continuing to the next exercise now, leave ArcMap open; otherwise, exit ArcMap.

Exercise 8c: Finish the map

The map still needs labels to identify prominent places and major roads. We'll label these features and convert the labels to a special kind of text called annotation that will give us complete manual control over placement. We also need to draw leader lines connecting the inset maps to their corresponding sites on the main map. Finally, we'll add a title, scale bar, north arrow, and acknowledgments.

1) Add and symbolize cities. To put some recognizable place names on the map, we'll add and label the *Cities* layer while making the features themselves invisible.

A If necessary, start ArcMap and open Lesson8.mxd from the list of recent map documents.

B In the Table of Contents, if necessary, activate the Main Map data frame.

C Open the data frame properties for Main Map and click the General tab, if necessary.

D Near the bottom of the dialog box, set the label engine to Maplex Label Engine and click OK.

E From the Catalog window, under ReadyData.gdb, add *Cities* to the layout.

In the Table of Contents, the layer should appear directly under the *LARiver* layer, which is where we want it.

F In the Table of Contents, rename the *Cities* layer to `City Labels`.

▷ How? Click the layer name to highlight it. Click again to make it editable.

G Open the Symbol Selector for the *City Labels* layer.

H In the Symbol Selector, set the Fill Color to No Color and set the Outline Color to No Color. Click OK.

Now the city features are invisible on the map.

2) Label cities. So far, we've done all our labeling from the Layer Properties dialog box. We can also use the Labeling toolbar, which is more convenient in some ways—for one thing, it lets you label all layers in the data frame from the same dialog box.

A Add the Labeling toolbar (Figure 8-71).

Figure 8-71

▷ How? From the main menu, choose Customize → Toolbars → Labeling.

B On the Labeling toolbar, click the Label Manager button.

In the Label Manager, each layer in the data frame is listed in the box on the left. A check mark indicates whether labels are turned on. (Right now, only *RecommendedSites* is labeled.) Underneath each layer name is a label class named Default. In the Default label class, all features in the layer are labeled the same way. If you wanted to vary the labeling properties within a layer (for example, to label large and small cities with different type sizes), you would create additional label classes.

Labeling properties for a layer are set on the right side of the Label Manager and with other tools on the Labeling toolbar. The functionality is the same whether you use the Labeling toolbar or the Labels tab in the layer properties—how you work with labels is really a matter of preference.

C In the scrolling box of label classes on the left, check the box next to *City Labels*.

D Under *City Labels,* click its Default label class to highlight it.

The property settings on the right side of the Label Manager now apply to the *City Labels* layer.

E In the Text String area, make sure the Label Field is set to NAME.

F In the Text Symbol area, set the font size to 9 and make it bold.

G Set the color to Gray 40%.

H Check your settings against Figure 8-72 and click OK.

Figure 8-72

The city labels appear on the map. (The positions of your labels may not exactly match the example below.) You're likely to notice some problematic placements (circled in Figure 8-73), such as:

- Labels crossing freeways
- Labels partially covered by inset maps
- Labels too close to the edge of the layout

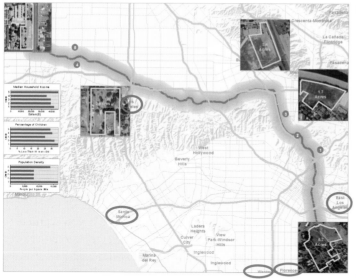

Figure 8-73

We also want to be selective about which labels are drawn. It's helpful to mark well-known places such as Beverly Hills and Santa Monica, but maybe not places that are less familiar (Ladera Heights, Westmont) or peripheral to our map (Pasadena). We also want to make sure that the label for Los Angeles is centrally placed.

We could resolve some of the problems by adjusting our label settings, but dynamic labeling will never give us complete control of placement. At this point, it makes sense to convert the labels into a type of text called map annotation that we can place at will. For more about labeling and annotation, see the topic *Labels, annotation, and graphic text*.

3) **Convert labels to map annotation.** We'll convert the city labels to map annotation. This is text that is stored in the map document, managed as a data frame property, and manipulated with tools on the Draw toolbar.

Ⓐ In the Table of Contents, right-click the *City Labels* layer and choose Convert Labels to Annotation.

Ⓑ In the Convert Labels to Annotation dialog box, in the Store Annotation area, click "In the map."

You can also store annotation as a geodatabase feature class, making it similar to point, line, and polygon feature classes. You would do this if you planned to reuse the annotation in other maps. In our case, the annotation serves the purposes of just this particular map, so it makes sense to store it in the map document.

Ⓒ In the Create Annotation For area, click "Features in current extent."

Ⓓ At the bottom of the dialog box, uncheck the box "Convert unplaced labels to unplaced annotation."

Ⓔ Check your settings against Figure 8-74 and click Convert.

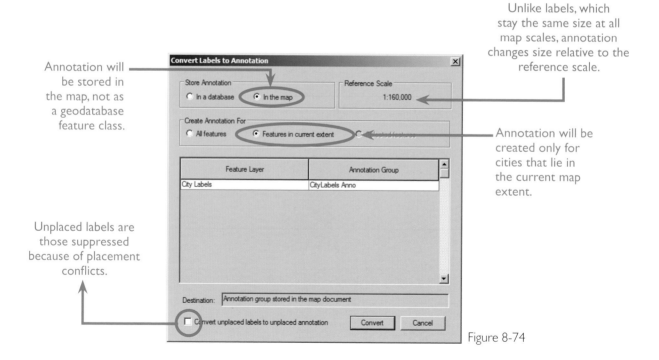

Annotation will be stored in the map, not as a geodatabase feature class.

Unlike labels, which stay the same size at all map scales, annotation changes size relative to the reference scale.

Annotation will be created only for cities that lie in the current map extent.

Unplaced labels are those suppressed because of placement conflicts.

Figure 8-74

Although the layout looks the same, something significant has happened behind the scenes. ArcMap placed annotation where the labels used to be and turned off dynamic labeling.

F Open the Main Map data frame properties and click the Annotation Groups tab (Figure 8-75).

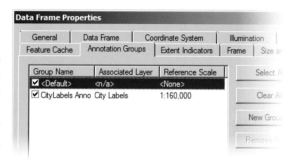

Every data frame has a <Default> annotation group, which holds text or graphics you create with the Draw toolbar in data view. (The black ellipses that marked your guesses in Lesson 1 were pieces of annotation stored in the <Default> group.) When you convert labels to annotation, a new group is created automatically. The annotation stored in different groups can be managed independently.

Figure 8-75

G Close the Data Frame Properties dialog box.

4) Edit the City Labels annotation group. We'll use the tools on the Draw toolbar to position or delete individual pieces of annotation.

A Add the Draw toolbar (Figure 8-76) and confirm that the Select Elements tool ▶ is selected.

Figure 8-76

At the moment, you can't select pieces of annotation, only layout elements such as the graphs or the data frames themselves. On a layout, two types of elements coexist: those that inhabit page space and those that inhabit map space. The locations of elements in page space are defined relative to the layout rulers. The locations of elements in map space—like our city annotation—are defined relative to a real-world coordinate system (in this case, State Plane California Zone 5). To work with elements in map space, you have to focus the data frame.

B On the Layout toolbar, click the Focus Data Frame button 🔳.

A thick, diagonal-line border appears around the data frame. We have left page space and entered map space.

C Click the Santa Monica piece of annotation to select it. Drag it away from the freeway, but keep it in the gray area that separates Santa Monica from Los Angeles (Figure 8-77).

Figure 8-77

Here, and in the rest of the step, modify the instructions as needed if your annotation isn't placed as shown in the figures.

The Los Angeles annotation is too far northwest. It's near the geographic center of the city (probably near your Site 4 data frame) but not near downtown, which is probably a more meaningful location. Malibu, squeezed in under the lowest graph, can be moved offshore.

D Click and drag the Los Angeles annotation toward downtown (think in terms of Dodger Stadium), as shown in Figure 8-78.

Move annotation to downtown area

Figure 8-78

E Click and drag the Malibu annotation down slightly into the water.

Some pieces of annotation may be partially covered by inset maps.

F Select the Florence annotation (lower right, by the Site 1 inset map) and press the Delete key on your keyboard.

G In the same way, delete the La Crescenta - Montrose annotation (upper right, by the Site 3 inset map).

H Move or delete other annotation, such as Burbank or Glendale, which is partially covered by an inset map.

There may be some pieces of annotation that are completely hidden behind inset maps. We could temporarily move the inset maps to find and delete (or reposition) these, but since they're not showing, we'll just leave them where they are.

Some other annotation is superfluous, identifying cities that aren't landmarks or that lie too far from the area of interest. Deciding which of this annotation to delete is subjective. On our map, we've deleted the following pieces (again, your annotation may not include all the same cities):

- Inglewood
- La Canada - Flintridge
- Ladera Heights
- Pasadena
- View Park - Windsor Hills
- Westmont

Some cities may be represented by two pieces of annotation. You can delete one or both pieces, as you like.

❶ Delete or reposition any other annotation to suit your cartographic sense.

▷ You can recover deleted annotation with the Undo button on the Standard toolbar if you need to.

In our map, we moved East Los Angeles slightly north to get it off a freeway, and we moved Glendale and Burbank away from inset maps that were partially covering them.

❶ When you're finished, click in some neutral part of the data frame (like the ocean) to unselect the last piece of annotation.

Your map should look more or less like Figure 8-79:

Figure 8-79

Figure 8-80

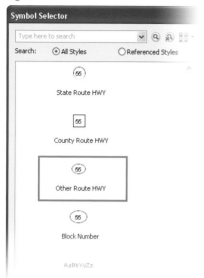

5) Place highway markers. We want to label the major roads, but not with their names, which would be hard to place legibly on our map. Instead, we'll use their numeric designations.

Ⓐ On the Labeling toolbar, click the Label Manager button .

Ⓑ In the list of label classes, check the box next to *MajorRoads*.

Ⓒ Under *MajorRoads,* click its Default label class to select it.

The property settings on the right now apply to the *MajorRoads* layer.

Ⓓ In the Text String area, set the Label Field to HWY_SYMBOL.

This field holds numeric descriptors for roads that have them.

Ⓔ In the Text Symbol area, click Symbol.

The Symbol Selector lists predefined text symbols, just like the predefined point, line, and polygon symbols you've seen earlier.

Ⓕ Scroll down and locate the Other Route HWY symbol near the bottom (Figure 8-80).

Ⓖ Click the symbol to select it and click OK.

Ⓗ In the Placement Properties area, confirm that Regular Placement is selected.

Ⓘ Click Position. In the Position Options dialog box, click Centered Horizontal and click OK.

Ⓙ Click Properties. In the Placement Properties dialog box, click the Label Density tab.

Ⓚ Check the Remove Duplicates box and click Options.

Ⓛ In the Duplicate Labels dialog box, for the Search radius, type **4** and set the units to Inches (Figure 8-81). Click OK on all open dialog boxes.

Figure 8-81

On the map, the major roads are labeled with text symbols (Figure 8-82). Duplicates are removed unless they're more than four inches apart. We could make slight placement improvements by converting the labels to annotation, but things are pretty good as they are.

Figure 8-82

6) **Add leader lines.** We placed the inset maps away from the site locations, so we need to connect them to their red symbols with leader lines.

Ⓐ If the data frame is still in focus, on the Layout toolbar, click the Focus Data Frame button 🖼 to remove the focus.

The diagonal-line border should be gone.

Ⓑ On the Layout toolbar, click the Zoom In tool. Zoom in on the Site 5 inset map and its corresponding symbol in the upper left corner.

Ⓒ On the Draw toolbar, click the Drawing drop-down menu and choose Default Symbol Properties.

Ⓓ In the Default Symbol Properties dialog box, click Line to open the Symbol Selector.

Ⓔ Set the Color to Mars Red and the Width to **0.5**.

Ⓕ Click OK on the Symbol Selector. Compare your settings to Figure 8-83 and click OK on the Default Symbol Properties dialog box.

Ⓖ On the Draw toolbar, click the drop-down arrow next to the Rectangle tool and choose Line (Figure 8-84).

Figure 8-83

Figure 8-84

H Click on top of the Site 5 symbol to start the line. Go straight up; then click to go left. Double-click on the edge of the inset map to complete the line (Figure 8-85).

Figure 8-85

Text or graphics you create with drawing tools when the data frame is *not* in focus—such as this leader line—are in page space and belong only to the layout. These elements won't appear in data view.

I On the Layout toolbar, click the Zoom Whole Page button.

J Click the Zoom In tool and zoom in on another inset map, such as Site 4, and its corresponding symbol.

K On the Draw toolbar, click the Line tool and draw a leader line connecting the symbol to the inset map.

Not all the leader lines need a right angle; some can be connected with a straight line. It's mostly a matter of what you think looks good. Refer to the finished map at the end of this lesson for ideas.

L Add leader lines for the remaining sites in the same way.

  ▷ If a line turns out badly, delete it and draw another one.

M When you're finished, click the Zoom Whole Page button and unselect the last leader line.

N Save the map.

7) **Add the map title**. We want the title to have the largest and most prominent typeface on the layout.

A From the main menu, choose Insert → Title to open the Insert Title box.

The title that we entered earlier in the map document properties appears automatically at the top of the layout.

Ⓑ Right-click the selected title and choose Properties.

In the Properties dialog box, on the Text tab, the title appears as a dynamic text tag (Figure 8-86).

Ⓒ Click Change Symbol.

Ⓓ In the Symbol Selector, make the Size 22 and the Style Bold. Click OK.

Ⓔ Click the Size and Position tab (Figure 8-87).

The X,Y position values are in inches, because the map title is in page space.

Ⓕ Click OK on the Properties dialog box.

Ⓖ Drag the title to the lower left corner of the layout.

This corner of ocean is empty space that we can use for the map title, north arrow, and scale bar. The title is too long, however. To put it on two lines, we have to convert it to graphics, which will break the dynamic text link.

Ⓗ Right-click the selected title and choose Convert to Graphics.

Ⓘ Right-click the title again and choose Properties. In the Properties dialog box, click the Text tab.

Ⓙ In the Text box, put the cursor between the words "River" and "Park." Delete the space between them and press Enter (Figure 8-88). Click OK.

Ⓚ Snap the text to the guide and place it over the ocean as shown in Figure 8-89.

Ⓛ Click outside the layout to unselect the text.

Figure 8-86

If you change the title in the Map Properties dialog box, this text will update.

Figure 8-87

The position of the title is specified in the layout's ruler units.

Figure 8-88

Figure 8-89

Figure 8-90

Draw splined text along this path with a few clicks.

Figure 8-91

Figure 8-92

If you switch to data view, you'll see the Pacific Ocean annotation, but not the map title.

8) Add Pacific Ocean annotation. Speaking of the ocean, we should identify it as the Pacific on the map. We don't have an ocean feature to label (the ocean is part of the *Terrain* basemap), so we need to create a piece of annotation.

Ⓐ On the Draw toolbar, click the Drawing menu, choose Active Annotation Target, and confirm that <Default> is checked.

We don't want to add this annotation to the CityLabels Anno group.

Ⓑ On the Layout toolbar, click the Focus Data Frame button 🖼 to put the Main Map data frame in focus.

Ⓒ On the Draw toolbar, click the Text **A** drop-down arrow and choose Splined Text ⁴⁶꜀.

With the Splined Text tool you define a curved path for your text before you type it. We'll draw a gentle curve more or less following the coastline between Santa Monica and Marina del Rey.

Ⓓ Click once to begin the text path, click again to define a midpoint, then double-click to finish the path (Figure 8-90).

When you double-click, an editable text box opens.

Ⓔ In the text box, type **Pacific Ocean** and press Enter.

The text is drawn on the curved path you typed.

Ⓕ Double-click the selected annotation to open its properties.

Ⓖ On the Text tab, click Change Symbol.

Ⓗ In the Symbol Selector, make the color Delft Blue (Figure 8-91).

Ⓘ Click the Font drop-down arrow and choose Palatino Linotype.

▷ If you don't have this font, leave the font set to Arial or choose something else.

Ⓙ Click the Size drop-down and choose 11.

Ⓚ Make the style Bold **B** and Italic **I**. Click OK on the Symbol Selector.

By convention, text for bodies of water is italicized.

Ⓛ On the Properties dialog box, click the Size and Position tab.

The X,Y Position values are in feet. This annotation belongs to map space—its position is defined relative to the origin of the California State Plane Zone 5 coordinate system.

Ⓜ Click OK on the Properties dialog box.

Ⓝ Drag the text to the approximate position shown in Figure 8-92.

Ⓞ Click in an empty part of the ocean to unselect the text.

Labels, annotation, and graphic text

Labels are text associated with map features. Labels are a layer property you can turn on or off; the label text is drawn from a field in the layer attribute table. Labels are dynamically placed by ArcMap, subject to preferences you set, and are repositioned as you zoom or pan the map. ArcMap has both a Standard Label Engine and a Maplex Label Engine with advanced placement capabilities.

OBJECTID *	Shape *	NAME	
1	Point	Avalon	City
2	Point	Attu Station	Census De
3	Point	Anahola	Census De
4	Point	Eleele	Census De

Labels come from this field.

Cities labeled with the Maplex Label Engine

Annotation is map text or graphics that you can place exactly where you want. It can describe map features or places on the map not represented by features: in general, it can be any text or graphics that belong to the body of the map. Annotation can be created piece-by-piece as you need it or by a batch conversion of labels. Converting labels to annotation is a common step in the later stages of map creation when you're ready to make final adjustments to text placement.

Annotation can be stored either in a map document or as a geodatabase feature class. How best to store it is mostly a question of reusability. If you plan to reuse the annotation, store it in a geodatabase; if not, store it in the map. Geodatabase annotation, like other feature classes, is edited in an edit session. Map annotation is edited with tools on the Draw toolbar.

Labels in the upper left map were converted to geodatabase annotation. Text strings and placement were then edited with annotation editing tools.

The *CitiesAnno* feature class can be added to any map document.

Graphic text includes the title, acknowledgments, and any marginal text that isn't part of the map body. It can only be added and seen in layout view. It can be placed outside or on top of a data frame, but either way, its position is defined in layout page units, not real-world coordinates. Graphic text is added with tools on the Draw toolbar while the data frame is not in focus. It can also be inserted as dynamic text which updates automatically.

The Pacific Ocean text is stored as map annotation in the <Default> group. The LA Airport text and symbol are stored in the Airports group.

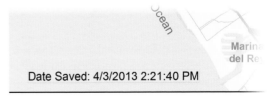

The Date Saved is dynamic text created from the Insert → Dynamic Text → Date Saved command.

Figure 8-93

Define
text path
here.

Figure 8-94

Figure 8-95

9) **Add river buffer annotation.** We'll add one more piece of annotation to clarify what the LA River buffer represents.

Ⓐ Make sure the data frame is still in focus.

Ⓑ On the Draw toolbar, click the Splined Text tool.

Ⓒ On the layout, define a text path inside the river buffer, south of the river, between the 405 and 101 freeways (Figure 8-93).

▷ If you want, you can click more than once along the path.

Ⓓ In the text box, type **LA River 3/4 mile corridor** and press Enter.

Ⓔ Double-click the selected annotation to open its properties.

Ⓕ On the Text tab, click Change Symbol.

Ⓖ In the Symbol Selector, make the color Fir Green (Figure 8-94).

Ⓗ Leave the font set to Arial, but change the Size to 8.

Ⓘ Make the style Bold **B** . Click OK in the Symbol Selector and Properties dialog boxes.

Ⓙ Drag the text to the approximate position shown in Figure 8-95.

Ⓚ Click somewhere else in the data frame to unselect the text.

Ⓛ On the Layout toolbar, click the Focus Data Frame button to take the data frame out of focus.

10) **Add a scale bar to the main map.** A scale bar will give the map reader a sense of distance on the map.

Ⓐ From the main menu, choose Insert → Scale Bar to open the Scale Bar Selector.

In the box on the left, you can choose from a variety of scale bar types.

Ⓑ Click Single Division Scale Bar to select it (Figure 8-96).

This is a simple, easy-to-read scale bar.

Figure 8-96

C Click Properties.

D In the Scale Bar dialog box, make sure the Scale and Units tab is selected.

E In the Units area, set the Division Units to Miles, if necessary. Accept the other default settings on this tab (Figure 8-97).

F Click the Format Tab.

G In the Text area, change the Size to 9.

H In the Bar area, underneath the Text area, click Symbol to open the Symbol Selector.

I Set the Fill Color to Arctic White and the Outline Width to **0.5**.

J Click OK on all open dialog boxes.

Figure 8-97

The scale bar draws on the map, but it's longer than we want.

K Make the scale bar shorter by dragging one of its side selection handles.

When you release the mouse button, the distance value is adjusted automatically.

L Resize the scale bar until its value is 3 Miles.

M Drag the scale bar to the bottom of the layout and place it approximately as shown in Figure 8-98.

Figure 8-98

N Click outside the layout to unselect the scale bar.

Scale bars and scale text

Scale bars provide a visual indication of the size of features and distance between features on the map. A scale bar is a line or bar divided into parts and labeled with its ground length, usually in multiples of map units such as tens of kilometers or hundreds of miles. If the map is enlarged or reduced, the scale bar remains correct. You can also represent the scale of your map with scale text, which is a verbal expression of scale. Scale text may relate equivalent units ("one inch equals 2 miles") or be a representative fraction that holds for any units ("1:100,000"). Scale text will be wrong if the map is enlarged or reduced after it has been printed.

Figure 8-99

11) Add a north arrow. We'll add a north arrow to remove any ambiguity about directions.

Ⓐ From the main menu, choose Insert → North Arrow.

Ⓑ In the North Arrow Selector, scroll down and click ESRI North 15 to select it.

We need a slender north arrow to economize on space in this corner of the map, but feel free to choose one you like.

Ⓒ Click Properties.

Ⓓ In the North Arrow dialog box, set the Size to 60 (Figure 8-99).

Ⓔ Click OK on both open dialog boxes.

Ⓕ Place the north arrow approximately as shown in Figure 8-100.

Ⓖ Unselect the north arrow.

Figure 8-100

12) Add data attribution. It's important to acknowledge data sources. Credits for the Imagery and Terrain basemap layers are included in the service layer credits that we moved off the layout in Exercise 8b. We also want to credit the City of Los Angeles Public Works Department, which kindly provided much of the project data. Finally, we want the map reader to know that the inset maps are at different scales. We'll combine all this information in one text element.

Ⓐ On the Layout toolbar, click the Pan tool 🖐 and pan to the area where you put the service layer credits.

Ⓑ On the Tools toolbar, click the Select Elements tool and select the service layer credits.

Ⓒ Right-click the selected text and choose Convert to Graphics.

This breaks the dynamic text link, as it did with the map title earlier, which makes it easier to edit the text. It also puts the service layer credits back in the data frames, so we'll have to remove them again when we're done.

ⓓ Double-click the selected text box to open its properties.

ⓔ In the Properties dialog box, on the Text tab, click some white space in the text box to unselect the text (Figure 8-101).

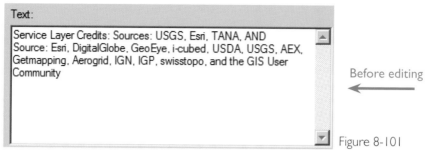

Before editing

Figure 8-101

The existing text includes credits from two basemap layers and has some redundancy. Don't take more pains than you want to, but the idea is to start the text with the words "Data sources:" and follow with the provider names separated by commas. Since the credits to USGS and Esri are repeated, you can delete one of the entries for each.

ⓕ Edit the text as described in the paragraph above.

ⓖ As the first data source in the list, insert **City of Los Angeles Department of Public Works – Bureau of Engineering**.

ⓗ Place the cursor at the end of the list. Press Enter twice to make a blank line.

ⓘ Type **Inset maps are at different scales**.

ⓙ Compare your text to Figure 8-102 and edit it to your satisfaction.

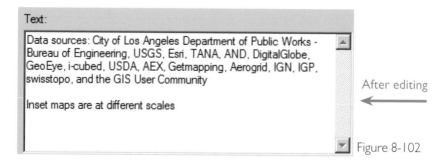

After editing

Figure 8-102

ⓚ Click Change Symbol. On the Symbol Selector, change the color to Gray 50%, the font to Arial Narrow, and the size to 7.

ⓛ Compare your settings to Figure 8-103. Click OK on the Symbol Selector and click OK on the Properties dialog box.

ⓜ On the layout, drag the text box to the lower left corner of the map.

Figure 8-103

N Optionally, zoom in on the layout with the Zoom In tool or the Zoom to Percent setting on the Layout toolbar.

▷ You may find a zoom level of 200% comfortable.

O Snap the text box to the guide; then resize and position the box to fit the space without crowding other elements.

P Click outside the layout page to unselect the text.

This part of the map should look more or less as shown in Figure 8-104.

Figure 8-104

Q From the main menu, choose Insert → Dynamic Text → Service Layer Credits.

The service layer credits are again placed in a text box. Your edited text is not affected in any way.

R Drag the service layer credits off the layout page.

As long as this box is outside the layout page, it won't print or otherwise intrude on the map.

S On the Layout toolbar, click the Zoom Whole Page button.

T Save the map.

In the next step, we'll preview and print the map. If you see any final adjustments you want to make, now is the time to do it. Look at the final map at the end of this lesson for reference.

13) Preview and print the map. If you have a printer available, you can print your map.

A From the main menu, choose File → Page and Print Setup.

B At the top of the Page and Print Setup dialog box, accept the default printer name or choose a different one.

Our map layout has a horizontal orientation, so we need the paper to match.

C In the Paper area, next to orientation, click Landscape.

D In the Map Page Size area, check the Use Printer Paper Settings box.

E Check your settings against Figure 8-105 and click OK.

Figure 8-105

The paper and map page are correctly oriented with respect to each other.

With the map size matched to the printer paper, a drop shadow is added to the layout. Your layout is now a preview of what you'll see when you print.

Figure 8-106

F If you see any problems, fix them.

G Save the map.

H On the Standard toolbar, click the Print button 🖨.

I On the Print dialog box, accept the defaults (Figure 8-106) and click OK to print your map.

J Optionally, remove the Draw, Labeling, and Layout toolbars from the interface.

K Exit ArcMap.

Hopefully, your map looks good. The next, and final, phase of the project is to share your analysis results online with a web map. You can see the lesson outline, and read a little bit about ArcGIS Online in the following pages; the lesson itself is a downloadable PDF file at the Understanding GIS web site.

Our goal in this book has been to guide you through the main steps of a GIS analysis problem: to explore the study area; to state the problem in quantitative terms; to choose, prepare, and edit the data; to plan and carry out the analysis; to automate the analysis with a model; and to present the analysis results. Not every GIS analysis problem is the same, but you should now understand the basic approach and have the requisite skills to take on the next project that comes your way.

Our second goal has been to introduce you to ArcGIS Desktop software. In planning this book, we made a decision to use the software only insofar as it served the needs of the project. We tried to make the project complex enough that we would have reason to use a lot of functionality; nevertheless, we definitely didn't see or do everything. Keep exploring the software on your own until the trial license period expires, and visit the book's web site at resources.arcgis.com/Understanding-GIS.

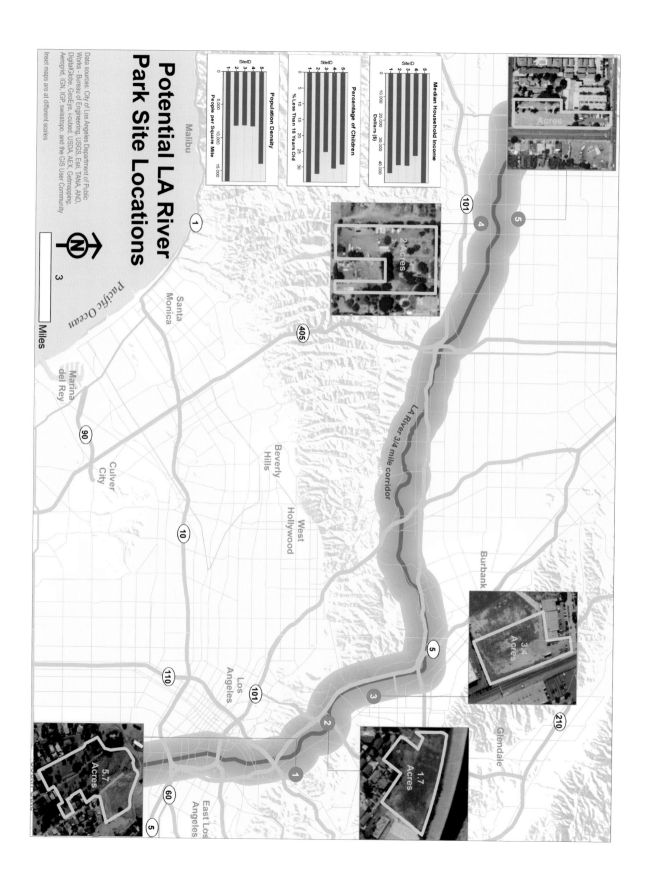

Potential LA River Park Site Locations

Median Household Income

SiteID

Dollars ($)

Percentage of Children

SiteID

% Less Than 18 Years Old

Population Density

SiteID

People per Square Mile

LA River 3/4 mile corridor

Data sources: City of Los Angeles Department of Public Works - Bureau of Engineering, USGS, Esri, TANA AND, DigitalGlobe, GeoEye, i-cubed, USDA, AEX, Getmapping, Aerogrid, IGN, IGP, swisstopo, and the GIS User Community

Inset maps are at different scales

N

0   3
Miles

Pacific Ocean

Malibu

Santa Monica

Marina del Rey

Culver City

Beverly Hills

West Hollywood

Los Angeles

East Los Angeles

Burbank

Glendale

2 Acres

3.4 Acres

1.7 Acres

5.7 Acres

Acres

Lesson

9 Share results online

THE CITY COUNCIL WOULD LIKE TO

make the park analysis available to the public. A paper map, such as the one we made in the last lesson, is the traditional way to share results, but online maps are a new way to share your story, methods, and data with the widest possible audience.

With a free personal account at ArcGIS Online, you can make web maps that combine Esri basemaps with your own geographic data and a huge number of map services (online layers) that are hosted on the web. You can share your web maps with everyone or with user groups that you set up. You can also publish the maps as web applications to give them a self-contained and professional appearance.

You can use your ArcGIS Online account to store many kinds of geographic items in addition to web maps. For example, you can upload geoprocessing packages, such as the one you made in Lesson 7, that contain models, tools, and data. Likewise, you can upload map and layer packages, which bundle map documents or individual layer files along with their source data. These packages can be shared from your ArcGIS Online account, allowing you to communicate your maps, results, and workflows to others. You can also store data in the form of shapefiles and tables in CSV format.

In the first exercise, we'll prepare relevant project data for use in the web map. This means converting the data from geodatabase to shapefile format, because shapefiles can be added to a web map. We'll also outline the steps for setting up your personal ArcGIS Online account.

We'll create the web map in the second and third exercises. In the fourth exercise, we'll publish the map as a web application. This allows users, and nonmappers in particular, to access the map as a self-contained web product (self-contained in much the same way as a PDF file) with a simple, attractive interface.

Finally, we'll upload our geoprocessing package and other files that we want to share at ArcGIS Online.

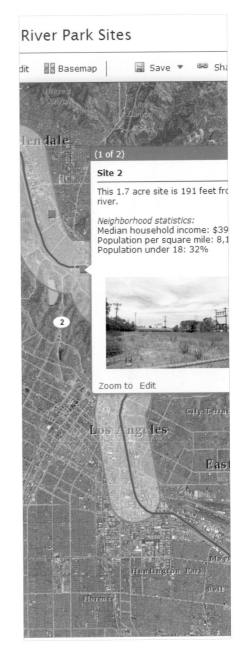

Lesson Nine roadmap

You are here:

What you'll do in this lesson:

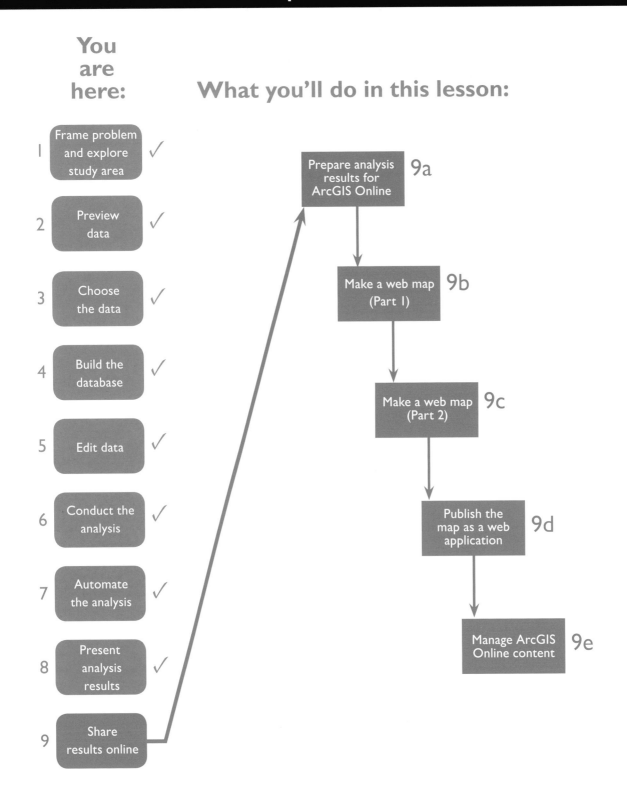

1	Frame problem and explore study area ✓
2	Preview data ✓
3	Choose the data ✓
4	Build the database ✓
5	Edit data ✓
6	Conduct the analysis ✓
7	Automate the analysis ✓
8	Present analysis results ✓
9	Share results online

9a Prepare analysis results for ArcGIS Online

9b Make a web map (Part 1)

9c Make a web map (Part 2)

9d Publish the map as a web application

9e Manage ArcGIS Online content

ArcGIS Online

Maps   Web Apps   Mobile Apps

Getting Ice Time

San Diego Shortlist

Global Development Atlas

Roots & Shoots: Sprouting a Greener Future

Esri Maps for Office 1.1

Landsat Series

IUCN Amazing Species

Roving the Red Planet

GeoStories

ArcGIS Online is GIS on the web. You can share your maps, data, applications, and workflows with everyone in the GIS community.

A selection of web applications that anyone can use from the growing ArcGIS Online gallery.

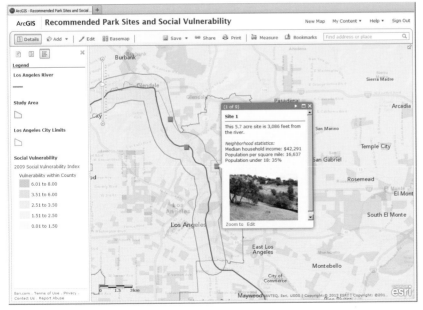

Create maps and applications from a web interface. Add your own data to Esri basemaps, and use online layers shared by Esri and others.

This web map shows the LA River park suitability results on top of a social vulnerability layer. The social vulnerability layer (comprising variables such as age, income, and disability) is a public map service at ArcGIS Online.

ArcGIS Online and ArcGIS for Desktop are part of the same system. Layers shared online can be added directly to ArcMap. Maps and analysis results created in ArcMap can be uploaded to ArcGIS Online.

In ArcMap, you can add layers from ArcGIS Online to your map document. These layers may be map services, which are served to your computer, or packages, which are downloaded and extracted.

My Content

		Title	Type	Modified	Shared	Size
□	☑	LA River Park Site Suitability Study	Web Map	Apr 8, 2013	Everyone	146 KB
□		LA River Park Study	Web Mapping Application	Apr 8, 2013	Everyone	1 KB
□	◇	LA River Proposed Park Sites	Map Service	Apr 8, 2013	Everyone	1 KB
□		LARiver	Shapefile	Apr 8, 2013	Everyone	38 KB
□		LARiverBuffer	Shapefile	Apr 8, 2013	Everyone	58 KB

The My Content page of your ArcGIS Online account is where you store, manage, and share items.

Items you can store include web maps, web applications, map services, data, and packages.

An ArcGIS Online subscription gives you even more capabilities. For example, you can publish your own map services and use online analysis and query tools.

This web map of suitable park sites includes a layer of facilities (red points) that release environmental toxins. An online analysis tool will count the number of toxic release locations in the buffer zone around the Los Angeles River.

Appendix

A Analysis issues

ONE THING YOU LEARNED FROM

the analysis is that the results you got—namely, the five sites recommended to the city council—depended on many different factors: the specific definitions you gave to the park siting guidelines in Lesson 2, the datasets you chose in Lesson 3, and several decisions you made in the course of data preparation and analysis. This is true for any analysis project: outcomes are determined by inputs, assumptions, and choices that are subject to reevaluation.

The scope and complexity of this project were somewhat limited by its context: after all, this is an introductory book about GIS analysis. In this appendix, we'll discuss a few issues generally relevant to analysis problems as well as some ways in which this analysis might be refined or expanded. We'll also touch on a few of its methodological soft spots. No GIS analysis is perfect or complete: being aware of limitations, and being candid about them, is a strength.

General considerations

Data quality

Good results require good data. Ideally, data should be accurate to a known standard, complete, up-to-date, and supported by metadata. We worked with published data from commercial vendors and operational data maintained by a large municipal agency: you can't do much better. Still, we saw concerns to address. For example, in Lesson 4 we loaded a couple of features that were missing from our parks dataset. In Lesson 5 we edited a feature with an inaccurate boundary and added a missing feature. Issues like these may not reflect negatively on the data. For one thing, large, comprehensive datasets, such as our parks dataset, are updated and published on schedules; they don't respond instantly to every change in the real world.

Also, data is generalized for use at particular scales. Our parks dataset was intended for use at scales of about 1:100,000. If you use data at a scale much larger or smaller than the scale it was designed for, or if your map mixes datasets that are meant for use at quite different scales, you're liable to encounter problems. Before criticizing the data, ask yourself if you're blaming a dog for not being a cat.

Data models

Our project relied on vector datasets (point, line, and polygon feature classes) for analysis. We used raster layers, in the form of basemaps, only for background display and context. In ArcGIS Desktop, a variety of geographic surfaces can be represented as raster datasets and analyzed with the extension products ArcGIS Spatial Analyst and ArcGIS 3D Analyst. These surfaces include elevation and its derivatives (slope and visibility), as well as soil, vegetation, temperature, rainfall, and many others. Had we incorporated a raster elevation surface and Spatial Analyst tools into our analysis, we might have considered slope as a park suitability factor. We could also have investigated questions such as whether or not the river is visible from a particular site.

In addition to raster datasets, ArcGIS supports other surface models. A Triangulated Irregular Network (TIN) represents a surface as an expanse of contiguous triangles of varying size and shape (Figure A-1). TINs are good for precision modeling of small areas and for calculations of surface area and volume. Terrain datasets and LAS datasets are models for rendering collections of lidar data, which may number anywhere from hundreds of thousands to billions of points, as continuous surfaces.

Figure A-1

A raster surface is composed of cells (pixels).

A TIN surface is composed of triangles.

The Modifiable Areal Unit Problem (MAUP)

The Modifiable Areal Unit Problem affects geographic analysis in general. The problem is that statistical data varies with the nature of the areal unit by which it is organized. An example from our analysis is shown in Figure A-2.

Figure A-2

The population density of the block group (hatched area) is 3,917 people per square mile—only half what it needs to be. The population density of the census tract (the entire polygon that includes the hatched area) is 9,071 people per square mile, which would meet the requirement.

The vacant parcel with a dark blue outline lies in a block group symbolized with a hatch pattern. It was rejected as a possible park site because the block group didn't meet the population density threshold. However, the census tract that contains the parcel *does* meet the density threshold. Had we used census tracts rather than block groups as our neighborhood unit—a valid choice—this parcel might have become a recommended site. (It might still have been disqualified on other grounds.) You can't really escape the Modifiable Areal Unit Problem. Had we used census tracts as our neighborhood units, we probably would have disqualified some parcels that satisfied requirements within their block group but not within the census tract.

Distance analysis

Our analysis of distances (to the river and to existing parks) was based on straight-line measurements: an oversimplification because people don't walk or drive in straight lines. Figure A-3 shows the situation with Recommended Site 1. The straight-line distance from the site to the river, represented by the black line, is about 0.6 miles, putting the site inside the 0.75-mile limit. But the street distance to the nearest river access point is 1.6 miles, as shown by the red line. The value of putting a park close to the river may not be defined entirely in terms of accessibility (for example, there might be value in seeing the river from the site), but accessibility is certainly the main point.

Figure A-3

A sophisticated analysis of distance involves not only streets but also barriers. Even if the black line in the figure followed a street, it wouldn't continue through the railroad yard on the river's eastern bank. Our analysis didn't account for barriers such as railroad yards or freeways that complicate or block access. (Likewise, it didn't account for shortcuts and other local knowledge people have about navigating in their neighborhood.)

To do complete network analysis, you need the Network Analyst extension for ArcGIS. This would allow you to create, for example, service area polygons showing the total street-accessible area around a site, either in terms of distance or travel time. Route-finding, as in the example on the previous page, can be done in ArcMap with no extensions.

Finding routes

On the Tools toolbar, click the Find Route button 🗺 to open the Find Route dialog box. On the Stops tab, you can add start and destination points by clicking on a map (any basemap, for example), by typing addresses, or by specifying a point layer in the map document.

Our use of straight-line distance has certainly understated true travel distances. Consequently, most of our recommended sites aren't as river-accessible as their distance values suggest. By the same token, the 0.25-mile value we used as a minimum distance from existing parks may imply that some areas have better park access than they really do.

Intended park use

One consideration missing from our guidelines was the intended use of the park. Although it's obviously a neighborhood-scale park, there are still distinctions to make; for example, is the park meant to support passive recreation (walking, picnicking, and casual play) or organized sports and activities? This question leads to others about the need for facilities, parking, street access, landscaping, and so on; these questions, in turn, raise questions about slope and park shape. A hilly site or an oddly shaped one (a string bean, for example) might meet our other requirements but be unsuitable from the point of view of intended use.

A quantitative analysis of slope requires ArcGIS Spatial Analyst (see the section on *Data models* in this appendix). Analyzing shapes to see whether one polygon fits inside another—for example, whether a baseball field will fit in a parcel—is a manual process that has to be evaluated with drawing, editing, and measurement tools. You can, however, use the Minimum Bounding Geometry tool to make a calculation of a shape's "regularity." The tool creates a feature class of minimum bounding shapes containing the input features (Figure A-4). (You can choose from various bounding shape options.) Dividing the area of the input feature by the area of the bounding geometry feature gives a regularity index—the closer the quotient is to 1, the more regular the shape.

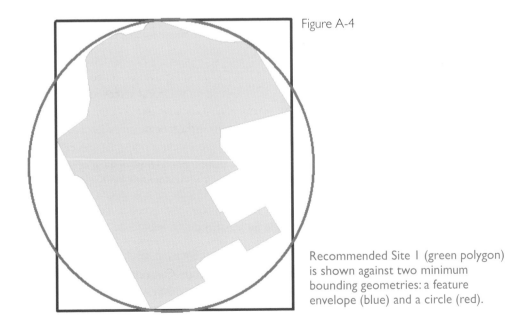

Figure A-4

Recommended Site 1 (green polygon) is shown against two minimum bounding geometries: a feature envelope (blue) and a circle (red).

Geoprocessing and spatial query on vacant parcels

Dissolving parcels

One of our guidelines was a minimum parcel size of one acre. In Lesson 4, we dissolved vacant parcels so we could consider the combined size of adjacent parcels that were individually smaller than an acre. This was useful for finding candidates (our five recommended sites are all composed of dissolved parcels), but it has some potential drawbacks.

Suppose we had both a minimum *and* a maximum size requirement. For example, what if our site had to be at least one acre but not larger than ten acres? Dissolving adjacent parcels might then disqualify some sites that were acceptable individually, like a five-acre site next to a six-acre site. This problem can be solved procedurally in ArcMap (that is, without custom scripts), but the steps are complicated.

A second potential problem has to do with how attributes are managed in overlay operations. One of the later steps in the analysis uses an Identity overlay to assign demographic values from features in *BlockGroups* to the five recommended sites. Suppose that one of the sites straddled two block groups: which block group's attribute values would the site get? (Normally, this wouldn't happen because individual parcels don't cross block group boundaries. Dissolving the parcels, however, makes it a possibility.) If it happened, the geoprocessing tool would arbitrarily choose one block group or the other and assign its values to the site. We might not want an arbitrary solution, however. We might want to use the values from the block group that shared the most area with the site, or the values that were most helpful to the analysis. Again, this problem can be solved procedurally, but not without effort. In our case, luckily, it didn't arise.

A third potential problem, inherent in any dissolve operation, is the loss of attributes in the feature class being dissolved. This wasn't an issue for us because we didn't need to maintain any parcel attributes except for ones, like ACRES, that we could calculate ourselves.

Selecting parcels by location

The first stage of the analysis focused on defining "good zones." These were areas that met our distance requirements with respect to the river and existing parks, and which also had suitable demographic values. After establishing these zones, we used a spatial query to select vacant parcels that fell within them. We chose the spatial operator "intersect" to select parcels that lay partially within a good zone; a containment operator would have selected only parcels that were entirely inside a good zone.

The main reason for our decision was simply to get more candidates. Our spatial query selected seven parcels (later reduced to five through aggregation). Had we used a "within" operator instead, four parcels would have been selected: the three that became Recommended Site 1, and the one that became Recommended Site 4. Sites 2 and 3 are partially inside the buffer zones of existing parks, and Site 5 is partially outside the LA River buffer: these three sites wouldn't have been selected by a containment operator. Figure A-5 shows Site 2 as an example.

Site 2 lies mostly in a light green "good zone" that meets all the analysis requirements. A small piece of it, however, crosses the quarter-mile buffer of an existing park.

We have another reason to use the "intersect" operator. Site 2, in the figure above, is a 1.7-acre site which was dissolved from seven adjacent vacant parcels. It so happens that five of those seven parcels can be combined to make a 1.3-acre site that lies entirely within the good zone (Figure A-6). Had we used a containment operator, Site 2 would not have been selected, even though a smaller desirable site is completely contained by the good zone at that location.

The other side of the coin with the "intersect" operator is that it selects a parcel even if only a small bit of it is within a good zone. This is the case, in fact, with Site 3.

Figure A-6

Site 2 consists of seven dissolved parcels. Five of these (selected) could have made a 1.3-acre site completely within a good zone.

Scoring

Our analysis had many siting requirements, but no way to rank candidates in terms of how well they met those requirements. We stated a preference for sites according to their proximity to the river and total population served, but these preferences weren't quantified in any way. The final evaluation of candidates may be too subjective and multifaceted to make strict scoring possible; nevertheless, we can briefly consider a couple of techniques.

For a variable of interest, you could map values onto a standardized score from zero to one by assigning 1 to the best value and 0 to the worst value and calculating other values proportionally. Using distance to the river as an example, you'd get the scores in the table below:

Scores are reversed because a low value is desirable.

Site	Distance to LA River (feet)	Standardized score (reversed)
1	3,086	0.15
2	191	1
3	2,482	0.33
4	2,537	0.31
5	3,603	0

Assuming all variables are equally important and can be standardized with the same linear function, you could rank the sites by scoring all variables and then summing the scores. (Ultimately, this approach needs normalization or else variables with small value ranges are differentiated as much as those with large value ranges.)

Your values may not be equally important, of course. A site's proximity to the river, for example, might matter more than the median household income in the neighborhood. This leads to the subject of multicriteria evaluation methods and the assignment of weights (multipliers) to variables. We won't pursue this subject here except to note that, in the raster analysis environment, ArcGIS provides a tool named Weighted Overlay to facilitate this kind of scoring.

Other factors

An analysis can be refined and developed to the extent that data, ingenuity, and time allow. In addition to the topics discussed above, we made simplifying assumptions and omitted several factors that might be important in choosing a park. For example:

- Land ownership. (We assumed that all vacant land was available. A refinement might consider only city-owned vacant parcels.)
- Land acquisition and development cost.
- Neighborhood crime rates.
- Environmental quality of the site. (This could include soil toxicity, proximity to freeways, and other health risks.)
- Accessibility from surface streets.
- Aesthetic qualities. (These might include visibility of the river, presence of trees, eyesores like overhead power lines, ambient noise levels, and so on.)

B Data and photo credits and data license agreement

Data and photo credits

Cover

Imagery sources: USGS, FAO, NPS, EPA, Esri, DeLorme, TomTom, other suppliers.

Lesson 1

Page 3 photo credit: Security Pacific National Bank Collection, Los Angeles Public Library, Anthony Friedkin, Photographer.

Imagery sources: Streets Basemap—Esri, DeLorme, AND, TomTom, First American, Esri Japan, UNEP-WCMC, USGS, METI, Esri Hong Kong, Esri Thailand, Procalculo Prosis; Imagery basemap—Esri, i-cubed, USDA FSA, USGS, AEX, GeoEye, Getmapping, Aerogrid, IGP.

Data sources: U.S. Populated Place Areas from Esri Data & Maps 2012, courtesy of TomTom, U.S. Census Bureau; River from Esri Data & Maps 2012, courtesy of USGS, Esri; U.S. and Canada Parks from Esri Data & Maps 2012, courtesy of TomTom; U.S. Tracts from Esri Data & Maps 2012, courtesy of TomTom, U.S. Census Bureau, Esri (Pop2010 field); U.S. Block Groups from Esri Data & Maps 2012, courtesy of TomTom, U.S. Census Bureau, Esri; LA City Land Parcels—courtesy of LA City, Public Works, Bureau of Engineering.

(Note: several values in the U.S. Block Groups data used throughout the book were altered for pedagogic reasons.)

Lesson 2

Imagery sources: Imagery basemap—Esri, i-cubed, USDA FSA, USGS, AEX, GeoEye, Getmapping, Aerogrid, IGP.

Data sources: LA City Land Parcels—courtesy of LA City, Public Works, Bureau of Engineering; U.S. Populated Place Points from Esri Data & Maps 2012, courtesy of U.S. Census Bureau; Parks.shp—courtesy of LA City, Public Works, Bureau of Engineering; U.S. Block Groups from Esri Data & Maps 2012, courtesy of TomTom, U.S. Census Bureau, Esri; U.S. Block Centroids—Esri Data & Maps 2012; U.S. Tracts from Esri Data & Maps 2012, courtesy of TomTom, U.S. Census Bureau, Esri (Pop2010 field); LA City Vacant Parcels table—courtesy of LA City, Public Works, Bureau of Engineering.

Lesson 3

Imagery sources: Physical Basemap—USGS, NPS, Esri, TomTom, AND; Imagery Basemap—Esri, i-cubed, USDA FSA, USGS, AEX, GeoEye, Getmapping, Aerogrid, IGP.

Data sources: LARiver Centerline—courtesy of LA City, Public Works, Bureau of Engineering; U.S. and Canada Water Polygons from Esri Data & Maps 2012, courtesy of TomTom; River from Esri Data & Maps 2012, courtesy of USGS, Esri; U.S. Populated Place Areas from Esri Data & Maps 2012, courtesy of TomTom, U.S. Census Bureau; Parks.shp—courtesy of LA City, Public Works, Bureau of Engineering; U.S. and Canada Parks from Esri Data & Maps 2012, courtesy of TomTom; VistaHermosaPark.tif—courtesy of Aerials Express from DataDoors; U.S. Counties from Esri Data & Maps 2012, courtesy of ArcUSA, U.S. Census Bureau, Esri (Pop2010 field); U.S. States from Esri Data & Maps 2012, courtesy of TomTom; Continents from Esri Data & Maps 2012, courtesy of ArcWorld Supplement; LA City Vacant Parcels table—courtesy of LA City, Public Works, Bureau of Engineering.

Lesson 4

Data sources: LA City Land parcels—courtesy of LA City, Public Works, Bureau of Engineering; U.S. Highways from Esri Data & Maps 2012, courtesy of TomTom; U.S. Counties from Esri Data & Maps 2012, courtesy of ArcUSA, U.S. Census Bureau, Esri (Pop2009 field); U.S. Populated Place Areas from Esri Data & Maps 2012, courtesy of TomTom, U.S. Census Bureau; U.S. and Canada Parks from Esri Data & Maps 2012, courtesy of TomTom; U.S. Block Groups from Esri Data & Maps 2012, courtesy of TomTom, U.S. Census Bureau, Esri; LA City Vacant Parcels table—courtesy of LA City, Public Works, Bureau of Engineering.

Lesson 5

Imagery sources: Imagery Basemap—Esri, i-cubed, USDA FSA, USGS, AEX, GeoEye, Getmapping, Aerogrid, IGP.

Data sources: U.S. and Canada Parks from Esri Data & Maps 2012, courtesy of TomTom.

Lesson 6

Site photos on pages 237–239 by Christian Harder.

Imagery sources: Topographic Basemap—USGS, FAO, NPS, EPA, Esri, DeLorme, TomTom, other suppliers; Imagery Basemap—Esri, i-cubed, USDA FSA, USGS, AEX, GeoEye, Getmapping, Aerogrid, IGP.

Data sources: River from Esri Data & Maps 2012, courtesy of USGS, Esri; U.S. Populated Place Points from Esri Data & Maps 2012, courtesy of U.S. Census Bureau; U.S. Counties from Esri

Data & Maps 2012, courtesy of ArcUSA, U.S. Census Bureau, Esri (Pop2009 field); LARiver Centerline—courtesy of LA City, Public Works, Bureau of Engineering; U.S. and Canada Parks from Esri Data & Maps 2012, courtesy of TomTom; LA City Land parcels— courtesy of LA City, Public Works, Bureau of Engineering; U.S. Block Groups from Esri Data & Maps 2012, courtesy of TomTom, U.S. Census Bureau, Esri; U.S. Block Centroids—Esri Data & Maps 2012; U.S. Large Area Landmarks from Esri Data & Maps 2012, courtesy of TomTom; LA City Vacant Parcels table—courtesy of LA City, Public Works, Bureau of Engineering.

Lesson 7

Data sources: LARiver Centerline—courtesy of LA City, Public Works, Bureau of Engineering; LA City Vacant Parcels table— courtesy of LA City, Public Works, Bureau of Engineering; U.S. Block Centroids—Esri Data & Maps 2012; U.S. Block Groups from Esri Data & Maps 2012, courtesy of TomTom, U.S. Census Bureau, Esri; U.S. and Canada Parks from Esri Data & Maps 2012, courtesy of TomTom.

The pancake recipe is thanks to Prof. Emeritus, Dr. Tech. Ole Jacobi, Institute of Surveying and Photogrammetry, Technical University of Denmark.

Lesson 8

Imagery sources: Imagery Basemap—Esri, i-cubed, USDA FSA, USGS, AEX, GeoEye, Getmapping, Aerogrid, IGP; Terrain Basemap—USGS, Esri, TomTom, AND.

Data sources: LARiver Centerline—courtesy of LA City, Public Works, Bureau of Engineering; LA City Land parcels—LA City, Public Works, Bureau of Engineering; U.S. Populated Place Areas from Esri Data & Maps 2012, courtesy of TomTom, U.S. Census Bureau; U.S. Counties from Esri Data & Maps 2012, courtesy of ArcUSA, U.S. Census Bureau, Esri; U.S. Highways from Esri Data & Maps 2012, courtesy of TomTom.

Lesson 9 (online)

Imagery sources: Imagery Basemap—Esri, i-cubed, USDA FSA, USGS, AEX, GeoEye, Getmapping, Aerogrid, IGP.

Data Sources: River from Esri Data & Maps 2012, courtesy of USGS, Esri; U.S. Median Household Income: © 2013 Esri.

Appendix A

Imagery sources: Imagery Basemap—Esri, i-cubed, USDA FSA, USGS, AEX, GeoEye, Getmapping, Aerogrid, IGP.

Data sources: U.S. Block Groups from Esri Data & Maps 2012— courtesy of TomTom, U.S. Census Bureau, Esri; U.S. Tracts from Esri Data & Maps 2012—courtesy of TomTom, U.S. Census Bureau, Esri.

Data license agreement

Important: Read carefully before opening the sealed media package

Environmental Systems Research Institute, Inc. (Esri), is willing to license the enclosed data and related materials to you only upon the condition that you accept all of the terms and conditions contained in this license agreement. Please read the terms and conditions carefully before opening the sealed media package. By opening the sealed media package, you are indicating your acceptance of the Esri License Agreement. If you do not agree to the terms and conditions as stated, then Esri is unwilling to license the data and related materials to you. In such event, you should return the media package with the seal unbroken and all other components to Esri.

Esri License Agreement

This is a license agreement, and not an agreement for sale, between you (Licensee) and Environmental Systems Research Institute, Inc. (Esri). This Esri License Agreement (Agreement) gives Licensee certain limited rights to use the data and related materials (Data and Related Materials). All rights not specifically granted in this Agreement are reserved to Esri and its Licensors.

Reservation of Ownership and Grant of License:

Esri and its Licensors retain exclusive rights, title, and ownership to the copy of the Data and Related Materials licensed under this Agreement and, hereby, grant to Licensee a personal, nonexclusive, nontransferable, royalty-free, worldwide license to use the Data and Related Materials based on the terms and conditions of this Agreement. Licensee agrees to use reasonable effort to protect the Data and Related Materials from unauthorized use, reproduction, distribution, or publication.

Proprietary Rights and Copyright:

Licensee acknowledges that the Data and Related Materials are proprietary and confidential property of Esri and its Licensors and are protected by United States copyright laws and applicable international copyright treaties and/or conventions.

Permitted Uses:

Licensee may install the Data and Related Materials onto permanent storage device(s) for Licensee's own internal use.

Licensee may make only one (1) copy of the original Data and Related Materials for archival purposes during the term of this Agreement unless the right to make additional copies is granted to Licensee in writing by Esri.

Licensee may internally use the Data and Related Materials provided by Esri for the stated purpose of GIS training and education.

Uses Not Permitted:

Licensee shall not sell, rent, lease, sublicense, lend, assign, time-share, or transfer, in whole or in part, or provide unlicensed Third Parties access to the Data and Related Materials or portions of the Data and Related Materials, any updates, or Licensee's rights under this Agreement.

Licensee shall not remove or obscure any copyright or trademark notices of Esri or its Licensors.

Term and Termination:

The license granted to Licensee by this Agreement shall commence upon the acceptance of this Agreement and shall continue until such time that Licensee elects in writing to discontinue use of the Data or Related Materials and terminates this Agreement. The Agreement shall automatically terminate without notice if Licensee fails to comply with any provision of this Agreement. Licensee shall then return to Esri the Data and Related Materials. The parties hereby agree that all provisions that operate to protect the rights of Esri and its Licensors shall remain in force should breach occur.

Disclaimer of Warranty:

The Data and Related Materials contained herein are provided "as-is," without warranty of any kind, either express or implied, including, but not limited to, the implied warranties of merchantability, fitness for a particular purpose, or noninfringement. Esri does not warrant that the Data and Related Materials will meet Licensee's needs or expectations, that the use of the Data and Related Materials will be uninterrupted, or that all nonconformities, defects, or errors can or will be corrected. Esri is not inviting reliance on the Data or Related Materials for commercial planning or analysis purposes, and Licensee should always check actual data.

Data Disclaimer:

The Data used herein has been derived from actual spatial or tabular information. In some cases, Esri has manipulated and applied certain assumptions, analyses, and opinions to the Data solely for educational training purposes. Assumptions, analyses, opinions applied, and actual outcomes may vary. Again, Esri is not inviting reliance on this Data, and the Licensee should always verify actual Data and exercise their own professional judgment when interpreting any outcomes.

Limitation of Liability:

Esri shall not be liable for direct, indirect, special, incidental, or consequential damages related to Licensee's use of the Data and Related Materials, even if Esri is advised of the possibility of such damage.

No Implied Waivers:

No failure or delay by Esri or its Licensors in enforcing any right or remedy under this Agreement shall be construed as a waiver of any future or other exercise of such right or remedy by Esri or its Licensors.

Order for Precedence:

Any conflict between the terms of this Agreement and any FAR, DFAR, purchase order, or other terms shall be resolved in favor of the terms expressed in this Agreement, subject to the government's minimum rights unless agreed otherwise.

Export Regulation:

Licensee acknowledges that this Agreement and the performance thereof are subject to compliance with any and all applicable United States laws, regulations, or orders relating to the export of data thereto. Licensee agrees to comply with all laws, regulations, and orders of the United States in regard to any export of such technical data.

Severability:

If any provision(s) of this Agreement shall be held to be invalid, illegal, or unenforceable by a court or other tribunal of competent jurisdiction, the validity, legality, and enforceability of the remaining provisions shall not in any way be affected or impaired thereby.

Governing Law:

This Agreement, entered into in the County of San Bernardino, shall be construed and enforced in accordance with and be governed by the laws of the United States of America and the State of California without reference to conflict of laws principles. The parties hereby consent to the personal jurisdiction of the courts of this county and waive their rights to change venue.

Entire Agreement:

The parties agree that this Agreement constitutes the sole and entire agreement of the parties as to the matter set forth herein and supersedes any previous agreements, understandings, and arrangements between the parties relating hereto.

Index

U

undoing operations 173, 181, 189, 278
unprojected data 108

V

vector data 66, 67, 350
vertices
  deleting 181
  described 185
  editing 179–184
  modifying 181–182
viewer windows 110–111

W

WGS 1984 107–108, 134
workflow. See analysis: planning
workspaces 192

Z

zooming
  to extents 33, 101, 117, 183
  to features 15, 99, 215
  interactively 9
  to layers 14
  in layout view 296, 332
  in models 261
  in previews 68

Topics and call-out boxes

The DVD included with this book
contains the project data only.

If the DVD is missing, visit:

resources.arcgis.com/understanding-gis

To download ArcGIS for Desktop trial software,
follow the instructions on page xiv.